Deconstructing LEGO

"In this insightful and engaging analysis of LEGO and its culture, Jonathan Rey Lee (de)constructs the 'brick' as a site teeming with cultural resonance. Examining the LEGO phenomenon through such interlocking perspectives as pedagogy, dramatism, digital culture, transmedia studies, and concepts of play, Lee's work embraces the building block mentality for scholars, fans, and AFOLs alike. Accessible and erudite, Lee proves he isn't just playing around."

—Paul Booth, *Professor, DePaul University, United States*

Jonathan Rey Lee

Deconstructing LEGO

The Medium and Messages of LEGO Play

palgrave
macmillan

Jonathan Rey Lee
Cascadia College
Bothell, WA, USA

University of Washington
Seattle, WA, USA

ISBN 978-3-030-53667-1 ISBN 978-3-030-53665-7 (eBook)
https://doi.org/10.1007/978-3-030-53665-7

Cover illustrations: SireAnko, Getty Images.

This Palgrave Macmillan imprint is published by the registered company Springer Nature Switzerland AG
The registered company address is: Gewerbestrasse 11, 6330 Cham, Switzerland

Preface—Deconstructing "LEGO"

"LEGO is not a toy," argues Finn's father in *The LEGO Movie*, "it is a sophisticated interlocking brick system" (2014). Toys, apparently, cannot be "sophisticated" without forcibly suppressing their playful elements (as the father attempts to do by regulating his son's participation and gluing the bricks together). Indeed, this counterintuitive denial that one of the world's best-known toys is truly a toy finds surprising resonances in scholarly discourse. "Strictly speaking, LEGO isn't a toy" argue the editors of *LEGO and Philosophy* in precisely this vein: "These little plastic bricks are more like a building material or medium, and probably have as much or more in common with bricks and paint than they have with most of the items in the toy aisle at the local megamart" (Bacharach and Cook 2017, p. 2). Underlying both rejections is an implicit claim of worthiness—that LEGO is worthy of adult hobbyism or philosophical attention[1] because it is too serious to be toyed with. While I agree that there is certainly value to analyzing LEGO as a medium, it is impossible to separate how LEGO functions as a medium from its status as a toy. By exploring its distinctive

[1] It should be noted that the purpose of the anthology is markedly different from that of this project. *LEGO and Philosophy* is part of the Blackwell Philosophy and Pop Culture Series and, like most books in the series, aims not to theorize a pop culture phenomenon but to leverage a pop culture phenomenon to make philosophical reflection more accessible. Consequently, the anthology has good reasons to consider the abstract idea of LEGO as separable from its actual status as a toy. This project, by contrast, aims to deconstruct the actual phenomenon of LEGO and cannot itself make any such abstraction.

synthesis of medium and toy, therefore, this project aims to deconstruct how LEGO's ideological and material design constructs its distinctive, paradoxical brand of playful yet serious, participatory yet consumerist, creative yet scripted play.

LEGO has undeniable cultural impact as an iconic multigenerational toy. In 1980, LEGO could be found in 70% of European households with children (Lipkowitz 2012, p. 24). In 2003, even while the company narrowly avoided bankruptcy, this statistic rose to 80% of North American and European households (Robertson 2013, p. 71). Since then, LEGO has only extended its cultural reach, becoming one of the three most recognizable global brands and taking the title of world's largest toy manufacturer (Robertson 2013, p. 3). Certainly, a popular culture phenomenon of this magnitude merits the critical attention that LEGO has only just begun to receive.

Yet, LEGO also necessitates more specific critical attention as a distinctive, boundary-blurring participatory media phenomenon. At once a *toy medium* (a meaning-making system) and *media toy* (a branded toy tied to its own and other media franchises), LEGO exemplifies the paradoxical intertwining of production and consumption that increasingly defines media culture. Consequently, deconstructing the medium and messages of LEGO both unravel the distinctive cultural contributions of this popular media phenomenon and provides a unique vantage point into the complex dynamics of an increasingly participatory media culture.

Fortunately, cultural critique of this influential participatory medium is gradually emerging in public consciousness. Anecdotally, by far the most common, immediate response to this project I have received (from scholars and non-scholars alike) is some variant on "it's so true that LEGO has become oversaturated with cultural messages—not at all like it used to be." While this demonstrates that the ideology of LEGO is becoming well-recognized, this critique targets the messages but not the medium of LEGO, as if reversing the more recent proliferation of socially constructed messages could restore LEGO to some originary neutral state. Yet, as this project will argue, there is no sense in which LEGO has ever been neutral or abstract.

This misreading is not entirely due to nostalgic misremembering, although that likely plays a part. Instead, this misreading is itself a cultural construction. LEGO is not abstract and therefore nonideological; rather, the idea of LEGO is constructed according to an ideology *of* abstraction.[2] Precisely because LEGO is ideological even at its most abstract, this construction toy invites *deconstruction*, critical interrogation into how its material design scripts play.

Certainly, what it means to deconstruct a toy—especially a "sophisticated interlocking brick system" like LEGO—differs from deconstructing other kinds of texts. Play systems differ from most traditional forms of media in that they are designed primarily to be possibility spaces for enacting various forms of player agency (physical manipulation, storytelling, etc.). In this way, toys are less narratives unto themselves and more conditions of possibility for emergent narratives. This means that toys do not fit neatly into the theories and methods of the linguistically focused Derridean school of deconstruction evoked by the very use of this term. For the sake of clarity, therefore, it is important to explore some of the ways this project does and does not overlap with this critical school whose name it freely redeploys.

In popular discourse, the theory of deconstruction is sometimes described as claiming the essential meaninglessness of language. This is not quite right. Instead, it is more accurate to say that deconstruction claims the essential *constructedness* of language. In other words, it argues that language forges new social meanings rather than merely naming absolute, objective meanings. This does not mean that there is no "true" way the world is—there is certainly some "objective" way things like rocks and gravity exist apart from human perception. Nonetheless, the deconstructionist points out that our understandings of things like rocks and gravity are built of much more than the things themselves—they are built, at least in part, of the discourses which circulate around them. Thus, what deconstruction denies is not truth itself but that humans can encounter truth in the abstract, unmediated by human considerations like culture, language, and perception.

Furthermore, deconstruction is primarily motivated by the ethical necessity of challenging restrictive or oppressive ways of thinking that

[2] The idealization of abstraction runs throughout the history of Western thought, becoming entwined with several ideological narratives pertaining to childhood. For instance, everyday discourse often implies that childhood is a space of innocence, that the play of the past was more natural before the intrusion of modern consumer culture, and that educational toys are developmental rather than socializing.

leverage this problematic idealization of conceptual abstraction to rationalize inequitable power relations. For instance, Derrida's deconstruction of language is informed by and directed toward the problematics of colonial language that he experienced as a monolingual French-Algerian forced to recognize that "I have only one language; it is not mine" (1998, p. 1). In this context, and in the way it is used here, deconstruction is less a philosophical claim about reality and more a tactic of critical resistance against the misuse of certain philosophical claims[3] about reality to rationalize or even enact social injustice. Critically, deconstruction calls into question prevailing conceptual systems because it cannot assume that these systems promote the universal good of all peoples.

Likewise, deconstructing LEGO matters because LEGO is also "not mine" for the vast majority of the world's population who are not white, middle-class boys. This is partially because not everyone has equal access to expensive LEGO toys, but more so because the ideological construction of LEGO presumes a very specific kind of subject.[4] Although this project does not primarily deconstruct these issues of access and presumed identity—this project aims instead to deconstruct the ideological formation of different modes of play—the possibility of material inequities being supported by ideological systems motivates any deconstructive project. LEGO matters both because it is culturally impactful and because its cultural impact is not equally distributed.

Importantly, the stakes of LEGO play are often quite subtle and nuanced. While any large multinational company leaves material impact by navigating the fraught ethical space of global capitalism,[5] I argue that even more impactful are the countless implicit, wordless promptings these toys weave into children's play. Despite and even because of its

[3] In addition to the Derridean model of deconstruction, my thinking in this regard is strongly influenced by the distinctive philosophical reflections of Ludwig Wittgenstein, who consistently challenges philosophical abstractions to reclaim more ordinary modes of understanding in ways that inform my practice of popular media analysis.

[4] This is not just an inference based on the representational politics of published designs; there is direct evidence that the LEGO company aims its designs at certain archetypal consumers (Landay 2014, p. 74); see also Chapter 3.

[5] While it is beyond the scope of this project, tracing the global production, circulation, and localization of LEGO products would be an excellent avenue for further research. One excellent example of this is Ashley Hinck's (2019) rhetorical analysis of the Greenpeace campaign to challenge the sourcing of LEGO plastics.

creative orientation, this ideologically laden toy invites children to become complicit in the ideological formation of their own ideological formation by providing carefully designed tools for mediating their imaginative, exploratory play. And while any individual moment of play will likely leave little material trace on society, the cumulative impact of the ideological formations woven into LEGO play may resonate much further than one might expect.

In this vein, this book is less an attempt to trace the material consequences of LEGO toys or LEGO play and more an attempt to interrogate the processes of ideological formation implicit in the scriptive design of LEGO play. Like the colonialism that inspired the development of deconstruction as a method of ideological resistance, the profit-driven capitalist system that frames children's media is ethically problematic even in its best-case scenario of a corporation being largely benevolent. While I believe LEGO has incredible potential for promoting creativity, being a product of a capitalist system means that LEGO is necessarily branded and commodified in ways that ideologically inflect the kinds of creativity it promotes.

While it is fair to say that a company as successful as LEGO must be doing something right, it is also vital to remember that few if any media are as inextricable from a single brand as LEGO. Whereas one can analyze the medium of painting apart from any particular brand of paints, there is no possibility of imagining the proprietary LEGO medium apart from the LEGO brand. Ethical questions necessarily arise—not because the company is ill-intentioned[6] but simply because its ideologies are often formed uncritically in a crucible of consumer demands and corporate pressures not well suited to critical design. This book, therefore, maintains the capitalization of "LEGO" used by the LEGO company and fan communities; not out of any affiliation with the brand, but instead as a persistent

[6]This project does not aim to make any particular claims about the ethical intentions of any past or present members of LEGO. Following what has become accepted wisdom in literary studies, this project sets aside questions of ***authorial intent*** to instead deconstruct the ideological formations implicit in the texts themselves. These formations may or may not have been directly intended by their creators. Thus, while I believe it is my responsibility as a media scholar to question corporately authored media, I have no reason whatsoever to think that LEGO is anything less than well-intentioned (especially in comparison to some other corporations).

reminder that LEGO is a fundamentally corporate medium. An abstract, ideologically neutral, noncorporate "lego" does not exist.[7]

To interrogate this always already meaning-laden toy medium and media toy, this book explores how LEGO play depends upon a paradoxically scripted creativity that raises ethical and ideological questions about participatory media and play. To accomplish this, this project primarily performs media-specific deconstructive analyses that unpack the ideological content of LEGO designs, arguably the most fundamental yet least theorized dimension of a medium whose scholarly interest has often been tied to its being "more than a toy."

Contextualizing the following analyses, this Preface traces the foundational ideological construction from the resonances of its blocks and bricks to its core values of *development, imagination, creativity* to some of the cultural entanglements that make deconstructing LEGO ethically pressing. Building on this foundation, Chapter 1 develops a conceptual framework for deconstructing LEGO by offering theoretical and methodological reflections on LEGO as a medium of *bricolage*. Then, the remainder of this project traces five ideologically rich forms of play—construction play, dramatic play, digital play, transmedia play, and attachment play—woven in and around LEGO toys and media.

Blocks and Bricks

Strip away the ideological noise of generations of increasingly complex LEGO toys and media, and what is left? A simple plastic brick. Yet, despite its humble appearances, this simple brick was never neutral. Indeed, this is precisely where the ideological construction of LEGO begins: in the space between the interrelated yet conflicting traditions of traditional abstract *blocks* and modern architectural *bricks*.

Tradition and modernization are also interwoven into the history of the LEGO company, a multigenerational family business that became a global megacorporation by abandoning its roots in wooden toymaking to build a brand around a mass-produced plastic brick. One might say that LEGO was created by technologizing and systematizing traditional building blocks to transform them into bricks. This transformation reflects

[7] Unbranded or DIY building blocks do exist but are not "LEGO" even if they sometimes work with the LEGO system.

the conflicting postwar impulses within which LEGO emerges, in which a nostalgic desire to return to an age of innocence existed alongside a seemingly incompatible progressive desire to celebrate technological advancement. An ideological as well as a material transformation, LEGO fuses the ideologies of block and brick into a paradoxical philosophy of traditional yet modern, nostalgic yet progressive, abstract yet representational play.

To better understand this complex ideological formation, some brief historical context on these two play traditions is in order. While building blocks have existed for millennia, the notion of blocks as explicitly educational toys was popularized by Enlightenment philosopher John Locke,[8] who is responsible for the wooden letter blocks still sold today. Although LEGO did produce similar letter blocks in 1946 (Lipkowitz 2012, p. 12), its philosophy of play is much more closely aligned with the theories of mid-nineteenth century educator Friedrich Froebel, best known for pioneering the kindergarten[9] system.

Whereas Locke believed that learning could be achieved by osmosis—that mere exposure to the alphabet would improve literacy—Froebel privileged the *activity* of play—more specifically, partially self-directed play within regulated pedagogical contexts. To this end, Froebel designed a series of toys known as "Froebel's gifts,"[10] "simple playthings, almost devoid of a local cultural context [that] were the symbols of a highly integrated system of learning that saw self-development socialisation and exploration of the environment as complementary facets of the growth of human knowledge" (Brewer 1980, pp. 38–39). Building on these educational philosophies, simple geometrical forms gained a privileged position in modernist developmental ideologies.[11]

[8] Locke argued that children learn to engage their world through play and are, therefore, particularly sensitive to the environments in which they play. By inscribing the alphabet upon traditional blocks, Locke hoped to familiarize children with language at an early stage. Although the effectiveness of this method can be questioned, Locke's theory was influential to the general understanding of child's play as productive that only increased in later centuries (Brewer 1980).

[9] LEGO began developing products for use in kindergarten classrooms in the 1950s (Robertson 2013, pp. 49–50).

[10] The idea of the gift was central to his theory because it meant that such play was presented as fun rather than compulsory, a line of reasoning that bears some similarity to contemporary discourses on gamification.

[11] Roughly speaking, *developmental play* refers to a common cultural history of considering play as a practical (or evolutionary) process of cognitive development. The history

This trend has only continued as block play has subsequently become a favored element in many Western visions of early child development. There is now extensive research demonstrating the cognitive benefits of block play, including several studies involving LEGO specifically.[12] Yet, the developmental benefits of such toys depend at least in part on how they are used—one study showed that building LEGO sets according to the instructions *reduced* creativity in subsequent tasks (Moreau and Engeset 2016). So, while LEGO certainly draws from this educational trajectory of block play, the cultural phenomenon of LEGO is clearly much more than a simple developmental tool.

While LEGO is loosely situated within this tradition of block play, LEGO's basic elements have always been *bricks* rather than blocks, drawing upon an architectural connotation is equally present in the Danish word "Mursten" originally used to name the bricks (Lauwaert 2009, p. 56). Although derivative of a more abstract construction toy,[13] the earliest LEGO toys were explicitly architectural (see Fig. 2.1). For instance, early designs of basic bricks contained now-absent slots "meant for the incorporation of doors and windows in LEGO constructions (that was the only play option these slots facilitated)" (Lauwaert 2009, p. 224) and early sets were released under a Town Plan (see Chapter 2). The resultant system is therefore more directly aligned with modern architectural construction toys[14] like Richter's Blocks, Lott's Bricks, and Bayko

and ideology of such developmental narratives are unpacked by Allison James and Alan Prout (2015) as they wrestle with the ethical challenges of applying various interpretive frameworks to children.

12 Gwen Dewar (2018) provides an accessible introduction to many of these studies.

13 LEGO toys were originally appropriated from Hilary Page's "Interlocking Building Cubes."

14 The association between LEGO and architecture is so strong it has gained traction even outside the world of toys. For instance, the booklet that accompanies the LEGO Architecture Studio set includes contributions from practicing architects who note ways they have used LEGO as a metaphor for construction and a tool for making architectural models. The booklet even notes that "There is a trend in current architecture fashion named "LEGO® architecture" because of its blocky and pixelated style" (2013, p. 78). There are also some possible comparisons to more industrial or mechanical building sets like Meccano and Erector Set that are explicitly designed and marketed as a way of introducing boys to engineering principles. This is especially true of the LEGO Technic line.

that emerged in the late nineteenth and early twentieth centuries as industrialization introduced cheap means of mass production used to provide products for a burgeoning childhood culture.

Although many of these toys are materially quite similar to traditional blocks, they are typically less abstract and more representational, placing more emphasis on the ideological content of represented designs. Consequently, modern architectural toys both draw upon the developmental ideology of the aforementioned history and offer unique ideological formations that merit specific deconstructive analyses like the ones in subsequent chapters.

At the same time, blocks and bricks have developed overlapping dispositions toward developmental and educational play, marshaling their materiality as tangible means of *grasping* relations or concepts. These traditions cannot be strictly separated historically, materially, or ideologically. Consequently, before turning to the more specific deconstructive analyses, this Preface further explores the distinctive philosophy of play that characterizes LEGO as a constitutive tension of block and brick, a doubled identity defined by fusing contradictory impulses to be *constructive*.

The LEGO Philosophy of Play

There can be no universal philosophy of LEGO play, as the medium is perpetually in flux as new elements and products are added to the system. Nonetheless, the avowed LEGO philosophy of play has remained surprisingly consistent over time—much more so than its products, whose design and marketing vary considerably across generations, target markets, and product lines. To contextualize an analysis of the material design of LEGO, therefore, it is vital to also deconstruct how the LEGO company cultivates an entire philosophy of play that attempts to script what its plastic products (should) mean.

A useful starting point for understanding LEGO's philosophy of play is the ten founding principles that have become a lynchpin of LEGO lore. As the story goes, Godtfred Kirk Christiansen, the successor to the family toy company responsible for developing the modern LEGO brick, was traveling via boat when he got into a discussion with a toy retailer who was lamenting the difficulty of selling toys that lacked a "system" to provide marketing continuity. Christiansen subsequently penned these principles to guide the development of a new toy system that eventually became LEGO. Given this motivation, it is unsurprising that three of these ten principles elaborate upon the idea of systematicity: "unlimited play potential," "the more LEGO, the greater the value," and "extra sets available." Also unsurprisingly, one principle essentially restates the motto from the wooden toy era of the company: "quality in every detail."

Most of the remaining principles aim less to define the qualities of the toy and more to define the ideal qualities of the play and players. Two principles state a desire to reach diverse target markets: "for girls and for boys" and "fun for every age." And three principles—"year-round play," "healthy, quiet play," and "long hours of play"—characterize the ideal play experience as domestic, indoor play. Together, these nine general principles articulate a clear ideal of play without yet specifying any specific design features. Here, an explicitly established philosophy of play precedes and informs toy design.

While these nine principles all left lasting marks on LEGO play, there is one more that I believe most clearly defines the toy: "Development, imagination, creativity." These three interwoven values comprehensively define the LEGO philosophy of play as a profound yet contested ideological intervention into childhood. Whereas most of the other nine principles are concrete goals for developing a commercially successful toy, this three-in-one principle provides a trinity of core *values* that name what childhood should be. Thus, while the other principles all refer either to the capacities of the toy or the nature of its play and players, this principle is the only one that refers to childhood itself—it is childhood, not LEGO toys, that is a space of *development, imagination, creativity*. Consequently, LEGO is defined first and foremost by the role it is designed to play in cultivating a specific ideological vision of childhood.

This philosophy reflects a modernist vision of childhood constituted by the collision of work and play. After all, "development" speaks to childhood as training for adulthood—that "play is the child's work." At the same time, "imagination" and "creativity" speak to childhood as a space of unbounded potentiality. In this seemingly paradoxical synthesis, imagination and creativity are both the means and ends of development—children practice imagination and creativity to develop into fully functioning adult members of a society that increasingly values its so-called *creative class*.

More than half a century later, LEGO continues to cultivate this singular developmental vision as its core brand identity:

> It is the LEGO® philosophy that "good quality play" enriches a child's life—and lays the foundation for later adult life. We believe that play is a key element in children's growth and *development*, and stimulates the *imagination*, the emergence of ideas, and *creative expression*. All LEGO products are based on this underlying philosophy of learning and development through play. (LEGO 2014, p. 3, emphasis mine)

Here, LEGO states its central vision, synthesizing *development, imagination, creativity* into a single, multifaceted ideal. Tellingly, this general ideology of play is sandwiched between two statements that refer to "LEGO products" and the "LEGO® philosophy." This neatly reveals the two primary ways LEGO cultivates play as *development, imagination, creativity*—through *toys* that function as tools for practicing these values and through a *brand* that comes to stand for these values.

As inheritors of the intertwined developmental traditions of blocks and bricks, LEGO toys already emerged ideologically and materially constructed to promote creative development. Building on these materialized cultural traces, the LEGO brand continues to actively cultivate an ideology of creativity fitted to LEGO play. As I argue elsewhere (2019), this includes sponsoring LEGO Foundation research reports that theorize the kinds of creativity most suited to systematic, toy-mediated play. More than merely academic, traces of this philosophy can be glimpsed throughout the history of LEGO design and marketing.

And this philosophy not only inflects actual LEGO products but also defines the brand, as indicated in how LEGO constructs its oft-repeated LEGO origin story, which Colin Fanning describes as "a case study in the selectivity of corporate history-telling" (2018, p. 90). One particularly telling retrospective is *The LEGO Story* (2012), a 17-minute animation depicting the founding of LEGO as a journey of *development, imagination, creativity*. Concealed amidst this retrospective's conventional celebration of hard work and persistence, moments of visual storytelling portray a particularly LEGO-like model of creativity being mediated by the material environment.

In one comic scene, LEGO company founder Ole Kirk Christiansen struggles to come up for a name for his fledgling company while scattered Locke-style letter blocks and the half-hidden words from a passing delivery truck scream "LEGO." In another scene, a frustrated Godtfred Kirk Christiansen is deep in thought working on a LEGO model of the first LEGOLAND theme park when an employee, not wanting to disturb him, places a new product next to the model. When Godtfred sees this serendipitously placed train set, he has an immediate "Eureka!" moment and integrates a train line into the park design. Here, LEGO products take a surprisingly agential role in mediating their own brand formation by doing precisely what the toys are advertised as doing—sparking creativity.

The moral of this story, eerily reminiscent of the LEGO Foundation research reports, is that creativity is much more materially and contextually grounded than the popular image of the free, spontaneous, intuitive imagination of a romantic genius. The LEGO vision of creativity suggests that creativity is best cultivated within the structuring influence of material systems like LEGO. More particularly, the LEGO vision of creativity closely resembles the material practice of *bricolage* (see Chapter 1), *the creative reassembly of already-significant elements*.

In this paradigm, the role of the toy is to provide a material system for this creative reassembly while the role of the brand is to advocate this creative paradigm. Thus, *The LEGO Story* narrates the origin of the toy as a natural outpouring of the values of the brand. Aligning the brand with the very ethos of creative reappropriation that sells its products, the retrospective even manages to acknowledge Godtfred's controversial appropriation of Hilary Page's existing plastic brick design while still celebrating his ingenuity in adapting it. The brick is born not of invention but of remix.

Together, toy and brand construct an ideology of play as *systematic creativity* (as the LEGO Foundation reports call it), the creative act of reassembling the already-significant material elements of the LEGO system within the already-significant ideological context cultivated by the LEGO brand. While there is no doubt that LEGO is a genuinely creative toy, its particular brand of creativity is heavily implicated in ideological constructions—not only the thematic content of various play themes like those analyzed in the following chapters but also in the ideological construction of play itself as a particularly LEGO-like vision of *development, imagination, creativity*.

Although there is certainly some merit to this strategy of establishing a carefully cultivated structuring context to facilitate creative expression,[15]

[15] One instance of this is the pedagogical concept of "scaffolding," which uses a construction metaphor to describe how learning or creativity can be engendered by learning scripts. The metaphor of the scaffold, a structure that is built as a place from which to build another structure, explains how a toy that was never abstract and has

such an ideologically laden medium cannot be considered neutral and abstract. Instead, LEGO facilitates ideologically laden play that demands deconstruction—including deconstructing the implication that such play can be neutral and abstract. Consider the subtle implications in this description of purportedly "abstract" sets of LEGO bricks:

> Bricks & More is the name given to sets or buckets with classic LEGO bricks and special parts such as windows, wheels, and roof tiles. No building instructions—just a bit of imagination. Run out of ideas? There are booklets enclosed—with illustrations to feed the active mind. (LEGO 2014, p. 4)

The harmoniousness of this statement relies upon a paradoxical blending of presence and absence. In the tradition of the block, LEGO advertises its "classic LEGO bricks" as supporting creative freedom by removing restrictions: "no building instructions—just a bit of imagination." Yet, in the same breath, it advertises representational content that acts rather like building instructions: "illustrations to feed the active mind." Imagination, it seems, wants to run free but needs to be "fed."

These seemingly contradictory statements are linked by a simple question: "Run out of ideas?" The positioning of this question implies that these illustrations are always available if needed but lie dormant otherwise. And while this is perhaps truer of Bricks ànd More (2009–present) than other more thematic LEGO sets, there are two serious flaws with this argument. First, as Jonathan Gray (2010) notes in his study of *paratexts* (see Chapter 1), consumers typically encounter the messages surrounding media objects before encountering the media texts themselves. It is implausible that these images could remain neutral at first glance and still be impactful when eventually turned to. Second, even if players ignore all paratextual elements, representational content is built into the medium itself through "special parts such as windows, wheels and roof tiles." LEGO was never designed to give free rein to the imagination—it was always designed as a medium of creative reassembly that explicitly cues players into certain modes of play.

become less abstract every generation can claim the general values of *development, imagination, creativity*. Yet, the role of scaffolding in creativity is determining not *how much* but rather *what kind* of creativity is encouraged. To scaffold, that is, is to shape as well as support.

Thus, it is particularly significant that LEGO rhetorically presents a demonstrably nonabstract material reality as abstract. Advertising its bricks as blocks, LEGO attempts to situate its brand of representational play within the developmental ideology of imaginative freedom. Rather than fleeing from ideology toward abstraction, LEGO cultivates an ideology *of* abstraction to advertise their representational designs as facilitating *development, imagination, creativity* and, perhaps more importantly, nothing more.

Cultural Constructions

As with most origin stories, this ideologically laden retelling of the past is actually about constructing a coherent identity in the present. Thus, *The LEGO Story* is just one piece of the much larger puzzle of how LEGO—one of the three most recognized global brands (Robertson 2013, p. 3)—cultivates its brand identity around the core values of *development, imagination, creativity*. Thus, this section further traces the cultural construction of this brand identity through several short case studies that reach beyond the more cultivated messaging of the LEGO origin story to consider how various noncorporate[16] communities[17]—children and their

[16]The LEGO brand is also co-constructed through collaborations with other corporate entities, such as licensed media franchises and toy retailers. These collaborations, moreover, do not always reinforce the LEGO messages. For instance, some toy retailers double down on LEGO's gendered targeted marketing (see Chapter 3) by dividing LEGO products across the pink and blue aisles while other retailers contradict this marketing by mixing all LEGO into a single display. Also, some of the more unique ideological aspects of *The LEGO Movie* films (see Chapter 6) may be attributable to filmmakers who worked with but not for LEGO. In the interests of clarity, I treat all of these "official" or "authorized" implementations of LEGO as part of a single larger web of interlocked corporate interests.

[17]The different interest-based affiliations of these communities all have their own identity politics that may or may not reflect LEGO's target market. LEGO design and marketing typically privileges certain narrow demographics: primarily young boys, secondarily young girls, and only thirdly adult fans (all implicitly presumed white and middle class). While it is beyond the scope of this project, it would be worth tracing how LEGO identity politics are responded to and reframed in moments of community uptake. For instance, AFOL communities often reframe a children's toy according to the identity politics of adult hobbyism. As Jennifer Garlen (2014) notes, AFOL communities are strikingly homogeneous: "Typically in their twenties and thirties, American AFOLs are most likely to be male, college-educated, and white. Older hobbyists in their forties and fifties are becoming more visible, however, as the fan community and Gen Xers age. Women

families, educational institutions, activists, artists, and adult fans—engage, perpetuate, and transform the ideology of *development, imagination, creativity*.

This deconstruction is a necessary first step to deconstructing the more media-specific ideologies of play discussed in later chapters. After all, the myth of abstraction that construes LEGO as neutral block play obscures any recognition of the ideological constructedness of LEGO toys. More subtly, even the myth that LEGO is abstract *underneath* its more ideological play themes obscures how ideological construction is embedded in LEGO down to its most essential material design. Unfortunately, many cultural responses to LEGO exhibit a problematic trend in which communities often deviate from or even reject outright the specific socializing messages of LEGO playsets in ways that subtly reinforce the myth of abstraction. While it is beyond the scope of this deconstructive project to conduct an extensive study of these cultural formations, these short surveys will shed light on the myth of abstraction to better contextualize the media-specific analyses of later chapters.

Ironically, by far the largest and most significant cultural context for LEGO play—the actual play of children—is the hardest to study, because children's play is ephemeral, imaginative, and often takes place outside adult surveillance (Giddings 2014, p. 241).[18] The imaginative activity of play cannot be captured except in idealizations. For instance, to visually convey imaginativeness, a series of LEGO "shadow" ads[19] depict extremely simplistic LEGO creations casting shadows of the real-world objects they represent, such as two bricks placed crosswise casting the shadow of an airplane. Yet, while such idealizations may poignantly represent the imagination, they cannot faithfully record actual play. Due to the

hobbyists are less common, especially in the ranks of the highest profile builders, but those who are active in the community are proud of their idiosyncratic interest and vocal in representing their segment of the overall group" (pp. 121–122).

[18] Despite these challenges, Giddings argues that we need further research into actual children's play with LEGO to avoid making unsubstantiated generalizations about its social impact. While I certainly agree, this project aims to complement rather than directly perform such sociological research. Deconstructing the medium and messages of LEGO provides insight into the systematic social forces LEGO exerts on children's culture. This primarily aims to contribute to more humanistic approaches to media studies but may also help generate hypotheses for future sociological research.

[19] This series of ads was developed by Blattner Brunner in 2006.

observer effect, most windows into actual children's play risk distorting how children actually play. Yet, because this project is a deconstructive analysis rather than a sociological investigation, a distorting window may still helpfully illustrate how LEGO play is culturally constructed by the normative cultural ideologies and discourses that surround play. After all, like much children's culture, LEGO is driven as much by how adults imagine children's play as by how children actually play.

For instance, one distorting window into actual children's play is photos of children with their LEGO creations published in the free LEGO fan magazines. Rather than providing a neutral window into a cultural phenomenon, these photos are twice curated—once by the children (and/or parents) who document and report their play, and again by the editorial staff who select and organize the photos. Consequently, while this feature provides some direct evidence of actual play, it primarily offers circumstantial evidence of what kinds of play children, parents, and LEGO executives want to publicize. As this private play is made public, it becomes impossible to disentangle the idealization of play from its represented reality.

In particular, the reality and ideal most often expressed in these photos show significant creative departures from retail playsets. As a quantitative analysis by Colin Fanning found, "only about one-third of published submissions directly mimicked the design language of existing LEGO products" (2018, p. 99). In other words, LEGO and its players often celebrate creative deviations from thematic playsets to cultivate an accepting, child-centric culture in which "A novice can stack bricks alongside the professionals and find acceptance" (Bender 2010, p. 49). Indeed, these images of children proudly displaying their creations are rhetorically presented as evidence of *development*, *imagination*, *creativity* as practiced by actual children and facilitated by the LEGO medium.

In a rhetorical move not uncommon in postmodern capitalism, the LEGO brand celebrates its own creative reinterpretation, thereby defining its brand as celebratory of its users' creativity. Consequently, the creative departures that LEGO celebrates are not genuine transgressions but rather creative extensions of the kinds of construction that LEGO actively facilitates. It is telling, for instance, that whenever a more abstract creation is depicted, it is much more likely to resemble classic LEGO construction than abstract modern art. Thus, while we are unlikely to ever have a perfect window into actual children's play, it is safe to say that most children's play neither wholly replicates nor wholly breaks from the ideological messages woven in and around their toys.

After all, ideologies are not static, compulsory dogmas. Ideologies are invitations[20] to participate in certain modes of collective thought and action. Consequently, ideologies thrive when implicitly perpetuated by creative individuals who adopt and adapt them. Similarly, the scripts that condition LEGO play do not exist solely within the confines of the explicit instructions but thrive whenever play resonates with the general or specific design philosophy of LEGO toys.

Like actual children's play, such ideological uptake is difficult to measure. However, one study found that even without explicit building instructions, paired builders produced cars that increasingly resembled each other as "each pair of participants seems to have consolidated their schematic representations of LEGO model cars, so that they became increasingly convinced what a LEGO car "ought" look like as they proceeded from one session to the next" (McGraw et al. 2014, p. 8). While this study more directly demonstrates the normalization of communal thinking, it is highly unlikely that the development of these "schematic representations of LEGO model cars" is completely independent from how LEGO toys are designed to construct cars. In the end, it is impossible to disentangle how material designs script LEGO play from the broader cultural ideologies pertaining to socializing yet creative play. And this is precisely the point—perhaps the main ideology that LEGO attempts to socialize children into is that of *development, imagination, creativity*. Socialization and creativity go hand in hand.

Thus, while much actual play with LEGO departs from the specific building instructions, such creative departures may also reinforce the underlying ideology of *development, imagination, creativity*. Another particularly telling instance of this can be found in the activist backlash against the problematically gendered LEGO Friends line (see Chapter 3). This backlash featured frequent citations to a 1981 print LEGO advertisement (see Fig. 3.1) depicting a young girl holding a hodgepodge creation like many featured in the Cool Creations page. After this ad went viral, journalist Lori Day interviewed Rachel Giordano, the woman who had

[20] Louis Althusser (1971) calls this the "hail"—the call that "interpellates" a subject into a subject of ideology.

modeled for the ad as a child. Reflecting on the changing faces of LEGO, Giordano remarks:

> In 1981, LEGOs were "Universal Building Sets" and that's exactly what they were…for boys and girls. Toys are supposed to foster creativity. But nowadays, it seems that a lot more toys already have messages built into them before a child even opens the pink or blue package. In 1981, LEGOs were simple and gender-neutral, and the creativity of the child produced the message. In 2014, it's the reverse: the toy delivers a message to the child, and this message is weirdly about gender. (Day 2014, n.p.)

Giordano is certainly correct that universal building was an ideal for LEGO in 1981 (although this was already starting to loosen as LEGO's two pioneering play themes—Space and Castle—had released three years prior). Expressing this ideological commitment, the very ad she starred in advertised such sets as designed to "help your children discover something very, very special: themselves." And she is certainly not alone in remembering LEGO as formerly more abstract, as similar rhetoric is easily found in activist posts and community forums.

At the same time, this ad undermines its own promise of abstraction. Noting that "Younger children build for fun," the ad continues to explain that "Older children build for realism" and that, therefore, these sets feature "more detailed pieces, like gears, rotors, and treaded tires for more realistic building." Similarly, while the featured set for younger children includes more gender-neutral builds like a suburban house, duck, and sailboat, the set for older children has much more thematic continuity, exclusively picturing yellow construction equipment. Clearly in 1981, LEGO was already delivering messages—more precisely, mixed messages that advertised abstraction and gender-neutrality[21] alongside representational, socializing content. Consequently, certain strains of consumer activism censure LEGO for its socializing content while calling for a return to an idealized neutrality that LEGO never possessed. These critiques, therefore, may only reinforce the myth of originary abstraction.

[21] Although the main tagline for this ad centers on the stereotypically feminized concept of the "beautiful," the ad itself presents a largely gender-neutral perspective. Neither Giordano's construction nor her outfit are visibly gendered. And, after the initial two lines which implicitly refer to the female Giordano ("Have you ever seen anything like it? Not just what she's made, but how proud it's made her."), the advertising copy instead refers to the more universal category "children."

In this cultural construction, the myth of abstraction is maintained in the very moment that LEGO's ideological construction is recognized and critiqued. This reflects a general trend in which the coherence of the LEGO brand identity is due not to eliminating mixed messages, but to embracing them—the prevailing LEGO rhetoric is defined by optimism that its constitutive contradictions are happily reconciled. LEGO thrives on being at once medium and toy, block and brick. As a hybrid of block and brick, LEGO facilitates multiple somewhat contradictory forms of play, exemplified by its use of contradictory labels like "serious play" and "hard fun"[22] to advertise its play.[23] Such oxymoronic slogans not only suggest that LEGO can simultaneously fulfill two contradictory demands, but rather *invite* users to transcend its surface messages in creative bricolage. Thus, while consumer activism typically advocates for LEGO to change its messages, other cultural practices actively remake LEGO into something more abstract.

Within the art world, for example, Brick Artist Nathan Sawaya has become known for drawing out the sculptural potential of the LEGO System (see also Chapter 6 Post-Script). Creating iconic sculptures from an extremely narrow palette of monochrome classic bricks,[24] Sawaya rejects the thematic messages of LEGO playsets and instead creates poignant pixelated forms that distil LEGO down to the simple elegance of the interlocking brick—a move that feels like a return to an essentially abstract origin but actually abstracts LEGO away from the significations that have always characterized the medium. As Sawaya explains,

> The LEGO brick also gives the viewer perspective. When someone looks at a sculpture built out of bricks they are going to be immediately struck by the distinct lines. Up close to the sculpture, one sees the plethora of rectangles, the many corners, the right angles. But when the viewer steps back and takes a look, they see it in a whole new way. All of those sharp

[22] The former derives from the work of Seymour Papert and names a LEGO method of cultivating creativity in professional environments; the latter derives from the comments of a young fan.

[23] Similarly, the famous advertising slogan "kid-tested, parent-approved" demonstrates how many child-centered products strive to simultaneously fulfill the pleasure-driven desires of the child and the development-driven desires of the adult caregiver.

[24] While Sawaya has many artworks in this style, most famous are a trilogy of human figures entitled *Yellow*, *Red*, and *Blue*. *Red* is discussed here and in the Chapter 6 Post-Script and *Yellow* is discussed in "The Plastic Art of LEGO" (Lee 2014).

> corners begin to blend together into curves. It is almost a metaphor of how people view art: it is all about perspective. Up close it may be simple rectangular bricks and corners, but from a different perspective, it's the human form and all of its curves. Further, it is made from a simple child's toy, but from a different perspective, it is contemporary art made from an accessible medium. It's been transformed from something quite ordinary—a toy—into something extraordinary—art. (2014, p. 213)

Building on this formal interplay of sharp corners and rounded curves, Sawaya's *Red* presents a humanoid figure who reaches toward the sky in a wordless cry while it either emerges from or dissolves into the pool of bricks at its waist. As a study on human life and/or death, this poignant image probes the horizon where being meets mere matter.

Furthermore, as a study on the LEGO medium, *Red* plays out the interplay of construction and deconstruction that defines the horizon of LEGO play. Seen as emerging from its component elements, this figure becomes a metaphor for the processes of bricolage (see Chapter 1) and digital assemblage (see Chapter 4) that characterize LEGO construction. Alternatively, when seen as dissolving into said elements, this figure becomes a metaphor for the atomistic yet plastic decomposition (Lee 2014) of LEGO. Either way, this art reconstructs LEGO as an essentially abstract substructure of pure form and connective potential, abstracting LEGO away from its surface messages to demonstrate what LEGO might look like as an artistic medium rather than a socializing toy.

Whereas the above examples mostly depart from LEGO's surface significations to draw out the more abstract aspects of the medium, a quite different approach to culturally reconstructing LEGO can be found in the creative production of the Adult Fan of LEGO (AFOL) community, which tends to be less interested in distilling the medium to an essential brick-based construction and more interested in cleverly repurposing its wide range of specific elements. This community, in other words, is generally known for another kind of resistant reading that flaunts creative and sometimes satirical integrations of symbolic elements. Like fan production in any media, these constructions do not merely reproduce the brand but reinterpret the brand, directly engaging its thematic elements to reassemble them into something new.

Whereas Fanning notes that the children's creations featured in the photo pages do not often resemble the LEGO design philosophy, AFOL creations (known as MOCs or "My Own Creations") often strive to *outdo* retail sets in scope, complexity, and clever part usage (Garlen 2014, p. 125), so much so that LEGO has an entire product line—LEGO Ideas (formerly Cuusoo)—dedicated to transforming AFOL creations into retail sets. Rather than using LEGO as abstract sculptural form, as Sawaya does, the best-known AFOL builds are intricate, richly textured constructions that push the boundaries of the construction system and/or cleverly reimagine the use of familiar parts. Thus, this manner of playful remixing still primarily reinforces the ideals of imagination and creativity (although it typically ignores developmental ideology as this community is definitionally not child-centric). Like much fan work, the creativity of AFOL bricolage typically transforms LEGO designs in ways that demonstrate a profound understanding (and appreciation) of those designs.

It is certainly no accident that so many cultural groups take up LEGO in ways that reinforce its designs or underlying ideologies. It is a testament to LEGO design that these communities seem to never tire of its vast and compelling possibility space (even consumer activists often frame their displeasure as feeling betrayed by a toy which they otherwise love). Yet, it is also a testament to the success of LEGO branding that many of the cultural constructions of LEGO dovetail with its underlying philosophy of play. While every generation has a toy fad that happens to be in the right place at the right time, it is safe to say that LEGO could not have achieved its unprecedented standing in the Western cultural imaginary without both strong toy design and effective brand formation.

To foster creative cultural appropriations that reinforce its core brand identity, LEGO actively engages the aforementioned groups in dialogues that extend well beyond the usual practices of advertising and social media presence.[25] To engage parents, LEGO publishes parenting resources such as its Whole Child Development Guide. To engage children, LEGO offers

[25] This approach has evolved over time. As David Robertson notes, "Less than two decades ago, LEGO was a fortress like company whose public position was "We don't accept unsolicited ideas." By 2006, the company had upended both the policy and its above-the-fray mind-set" (2013, p. 213). Describing the culture that resulted from this shift in mentality, he continues "LEGO came to realize that while open-source innovation can be managed, it can't be controlled. The process is best understood as an ongoing conversation between the company and its vast crowd of fans. Like any good dialogue, LEGO-style sourcing was built on the principles of mutual respect, each side's willingness to listen, a clear sense of what's in play and what's out of bounds, and a strong desire for mutually beneficial outcome. For outside collaborators, the reward could be intrinsic—such as recognition from peers and access to LEGO—as well as financial. As for LEGO,

the longstanding LEGO Club Magazine (now entitled LEGO Life and accompanied by an app) and interactive experiences in LEGO Stores, LEGOLAND theme parks, and LEGO Discovery Centers. To engage educational institutions, LEGO Education supports STEAM[26] learning initiatives, including Mindstorms robotics competitions. To engage other institutions, LEGO Serious Play uses toy construction to promote creative thinking in workplace environments. To engage artists, LEGO has offered select builders (including Sawaya) the opportunity to become LEGO Certified Professionals, who receive perks from the company in exchange for following certain community guidelines (although LEGO is also widely known for clashing with artists like Zbigniew Libera and Ai Weiwei who try to make political statements LEGO does not sanction). To engage AFOLs, LEGO maintains a Community Engagement team to specifically interface with fan communities and has at different times offered various ways of officially recognizing fans through programs like the LEGO Brand Ambassadors, LEGO User Groups, Recognized LEGO Fan Media, etc. And these are only a few notable examples of how LEGO engages the ongoing and evolving dialogues that contribute to its cultural construction.

Significantly, most of these examples represent *partnerships*—be they implied partnerships, as when parents work with the provided resources or explicit partnerships like the Certified Professionals program. In other words, LEGO cultivates its own play culture not only[27] by exerting regulatory pressure on these groups but also by positioning these groups as collaborators with or even co-creators of the LEGO brand. As Jonathan Bender notes, "It used to be that LEGO created value, but now value is being created across the community" (2010, p. 65). Thus, as with most ideological constructions, these relationships are thus founded more

the conversation almost certainly tightened its ties to the fan community. And in some instances, it delivered products that LEGO itself had never imagined" (pp. 213–214).

[26] LEGO uses this variant on STEM (Science, Technology, Engineering, Mathematics) that adds an A for "Art," presumably because LEGO wants to be seen as also facilitating artistic creativity even though most of its educational initiatives are more explicitly STEM than STEAM. For further analysis of LEGO's relationship to STEM education, and its association with cultures of whiteness, see Hinck (2019).

[27] Certainly, LEGO has had plenty of more litigious and acrimonious encounters, but I believe its ideological impact is much more subtle and effective in its more collaborative endeavors.

on complicity than direct control. Governed by the self-reflexive, self-fulfilling ideals of *development, imagination, creativity*, LEGO invites diverse communities to remake the medium in ways that often ultimately reinforce or even celebrate LEGO as a medium of *development, imagination, creativity*.

In sum, the cultural construction of LEGO is a blend of corporate design and advertising practices, public discourses that circulate through various social channels, and instances of private play. Explicitly co-produced media like the LEGO magazine photo pages and LEGO Ideas playsets only exemplify the deep reciprocity between corporate and community influences that necessarily characterizes the cultural construction of LEGO. While LEGO is most obviously a material medium for literally constructing things, LEGO also figuratively constructs and is constructed by these cultural value systems. While these cultural interplays are each easily worthy of study, here these dynamics provide the implicit (and occasionally explicit) context for the following deconstruction of official LEGO media texts and paratexts, ideological formations which gain meaning only within an assumed cultural context already shaped by the LEGO brand.

The Means and Ends of Deconstructing LEGO

Whereas cultural responses to LEGO often co-construct shared ideals of *development, imagination, creativity*, this critical project seeks neither to confirm nor deny whether LEGO achieves its core values. Nor does this project take a unilateral ethical position on the cultural impact of LEGO. To do so would presume a universal, static, deterministic cause-and-effect relationship between toys and social change, when in fact the meaningfulness of this interactive medium is constantly being remade. Instead, to ethically engage this dynamic medium, this project deconstructs key ideological forces at play in LEGO—not to censure their fixed meanings, but to critically intervene in the ongoing cultural formation of the medium.

In lieu of sweeping ethical claims, therefore, this project deconstructs ethical dynamics which largely take the form of constitutive tensions or paradoxes, mixed messages in which seemingly contradictory values intermingle to produce hybrid experiences. The most sweeping ethical generalization I can make is that this toy medium and media toy typically embraces the contradictions and paradoxes that define it. Rather than producing a single, coherent narrative, many ideological playscripts run through a diverse field of LEGO products, situating LEGO play within a complex network of ideological pushes and pulls. Consequently, deconstructing LEGO entails not only deconstructing foundational values like *development, imagination, creativity* but also deconstructing various other ideological threads that make up this network.

The project of deconstructing LEGO is particularly pressing because construction toys like LEGO are not commonly subject to ideological critique, especially in comparison to girls' toys like dolls. This critical bias, which may inadvertently reinforce the gendered inequalities such critiques are designed to combat, is unfortunately a common disposition in popular discourse: for instance, LEGO received more criticism for feminizing its LEGO Friends line than for a decades-long process of masculinizing its entire system.[28] This is an especially pernicious trend because it risks reinforcing underlying assumptions about what makes boys' toys "better," such as treating *development, imagination, creativity* as universal, gender-neutral values instead of the culturally constructed and often gendered values that they are.

The danger of this double standard is not only that it is inequitably gendered but also that it misleadingly exempts things like *development, imagination, creativity* from the realm of socialization. Socialization is not the uncreative, passive uptake of fixed ideologies; it is the performative, active process of developing cultural competences. Conversely, creativity is not the free and spontaneous generation of a newness; it is a dialogic and recombinative process that participates in the continual reconfiguration of culture. Feminizing the former and masculinizing the latter is symptomatic of a partial vision that obscures the creativity of girls' toys[29] and socialization of boys' toys. Deconstructing the ideologies at play in LEGO matters not because such ideological forces render LEGO play *un*creative,

[28] A notable exception is Anita Sarkeesian, who does an excellent job of rooting the former problem in the latter in her two-part critique of LEGO Friends on YouTube (2012).

[29] Barbie can be described as a construction toy, *a modular system for creatively constructing fashion assemblages.* After all, merely replacing the word "fashion" with "architectural" makes this description perfectly fit construction toys. That Barbie is never described this way, however, raises questions about the gendered cultural assumptions that make fashion frivolous and/or socializing and architecture educational.

but rather because such forces attempt to redirect creativity along predetermined pathways that reinforce prevailing cultural ideals and, at times, inequalities.

As the ethical aim of this project is to make multiple incursions within a vast and evolving field of play possibilities, the *methodological* aim of this project is to practice an experimental mode of media scholarship fitted to the distinctive blend of medium and message that characterizes LEGO. This is a far stranger and more wonderous task than I had initially thought. Imagine a theory of painting if the only painting equipment available throughout history were paint-by-numbers kits. Imagine a theory of poetry if the only medium for writing poetry throughout history were packets of refrigerator magnets with words on them. In the unfathomable world where most media operated like this, there would be no concept of a blank canvas for creative expression and no concept of a medium as a mere recording or transmission technology. Instead, this world might describe all its media as "some assembly required" but "content already included."

Strange though this may seem for traditional media, this is precisely how a patented commercial construction toy like LEGO operates. LEGO constitutes a complex system of meaning-making with incredible constructive potential that comes packaged in presorted kits consisting of preformed elements with a single brand marker etched on every stud. Deconstructing such toys, therefore, requires new paradigms not beholden to most established methods of textual analysis. To deconstruct a toy is to deconstruct the materiality that scripts its playful performances. To deconstruct a toy is to deconstruct a commercialized prop for *development, imagination, creativity*. And to deconstruct a toy is to deconstruct how it directs implicit ideological promises and invitations toward the playing subject.

To deconstruct this rather unusual medium, the theoretical and methodological provocations presented in Chapter 1 play with the notion of *bricolage*—a practice of fractious assemblages of scrounged elements that both describes LEGO and inspires this project to construct a media theory in a somewhat unusual way. Rather than presenting a comprehensive unifying field theory of LEGO, this project builds upon a series of theoretical gestures, piecing together a media theory scavenged from a diverse interdisciplinary array of theoretical concepts. The result will not be a singular theory but rather an assemblage of theorizations that shows its seams, rather like the visibly fractious assemblages constructed in LEGO. So, although these interdisciplinary touchstones often do not always fully cohere or directly address LEGO, together they provide a series of useful vantage points from which to triangulate the multifaceted LEGO experience.

Building on these provocations, the following chapters deconstruct how LEGO design participates in the ideological formation of five modes of toy play: *construction play*, *dramatic play*, *digital play*, *transmedia play*, and *attachment play*. To contextualize these analyses, each chapter draws on at least one primary scholarly discourse and invokes related trends in the cultural history of toys and play. Then, to conduct these analyses, each chapter deconstructs the ethical and ideological dynamics at play in a particular LEGO product line:

- Chapter 2, "Housing Play," draws upon the philosophy of architecture and the history of construction toys to explore *construction play*—the material bricolage of tangibly constructing LEGO structures—from the Town Plan to LEGO City. As the otherwise abstract processes of LEGO construction become inextricably tied to architectural significances, this chapter explores how miniature LEGO houses, towns, and cities embrace a suburban ethos that domesticates its construction play.
- Chapter 3, "Playing House," draws upon theories of gender performativity and the history of dolls, dollhouses, and toy theaters to explore *dramatic play*—the performative bricolage of playing out narratives with LEGO toys—from early thematic playsets to the "for girls" LEGO Friends product line. Beyond gendering individual play themes, the ideology that emerges in this chapter also genders the toy medium itself, masculinizing construction play and feminizing dramatic play. This ideology thereby bifurcates both the LEGO medium and its play. This chapter forms a dyad with Chapter 2 that explores how the tension between housing play and playing house defines and complicates the toy medium.
- Chapter 4, "Digital Analogs," draws upon theories of digitality to explore *digital play*—the self-referential bricolage of playing with assemblages of discrete elements—from material LEGO bricks to the virtual gameplay of *LEGO Worlds* and *LEGO Dimensions*. This chapter presents a conceptual frame to bridge the flanking dyads of LEGO as toy medium (Chapters 2 and 3) and media toy (Chapters 5 and 6). Showing how the digital LEGO idea transcends its material

and virtual incarnations, this chapter explores how the LEGO brand has come to rely on the constitutive interplay of digital and analog experience to present itself as a medium of bricolage.

- Chapter 5, "Story Toys," draws upon theories of transmedia storytelling and the history of character toys to explore *transmedia play*—the bricolage of mobilizing licensed toys to explore transmedia worlds—in LEGO *Star Wars*. This chapter delves into the paradoxical ways LEGO's story toys adopt a filmic logic that faithfully reproduces canon, even as its media paratexts adopt a toy-centric logic that playfully reimagines canon. Theorizing this paradox as a constitutive tension between *play* and *dis-play*, this chapter traces how the bricolage of transmedia play mobilizes fixed signifiers to simultaneously script narrative play and play with narrative scripts.
- Chapter 6, "Toy Stories," draws upon a language of attachment derived from psychological discourses and the tradition of toys-to-life narratives to explore *attachment play*—toy-mediated storytelling that expresses a need for emotional attachment—in *The LEGO Movie* and *The LEGO Movie 2*. Across four layers of filmic meaning—surface attachment quests, the imaginative storytelling of the child characters, animated toys-to-life narratives, and the object-agency of the toys themselves—LEGO brands itself as actively promoting attachment. Furthermore, by positioning the toys as both a metaphor for and active mediator of emotional attachment, LEGO constructs a problematically consumerist ethos of connectivity. As story toys and toy stories mutually construct each other, this chapter forms a dyad with Chapter 5 to explore the dynamics that shape the formation of LEGO as a multimedia and transmedia phenomenon.

Despite organizing this project around these five forms of play, the categories outlined above are neither absolute nor exhaustive. Instead, countless overlapping and evolving systems of meaning are woven together within the vast possibility space of LEGO play.

To point beyond the scripted messages of corporately cultivated LEGO play, six short Post-Scripts interlaced between these chapters explore exemplary LEGO artworks and fan creations that variously challenge LEGO paradigms. These Post-Scripts offer case studies of meaningful resistances or alternatives to the core ideological scripts discussed in the chapters, serving as important reminders that the possibilities for LEGO play always extend beyond the narrower forms of play implied in LEGO's ideologically laden playscripts. In addition, to further reflect on how the meaningfulness of this possibility space transcends articulation, the final section, "After Words," considers *sandbox play* as the mixing and meshing of multiple modes of play within a multifaceted play experience.

In an open field like the relatively uncharted geography of LEGO play, it is easier to justify the inclusions than the exclusions, not least because there are quite a few more of the latter than the former. The following chapters by no means exhaust the continually shifting geography of play, which spans hundreds of product lines and thousands of sets. Nor do they venture into the more or less compatible LEGO-brand building systems of Technic, Mindstorms, and Bionicle[30] or non-LEGO imitations like Mega Bloks or KRE-O. While any of these would be fertile ground for analysis, they are better positioned for a subsequent study since their significance in many ways respond to core LEGO play.

Even more difficult was the practical necessity of merely gesturing to or bypassing many interesting critical approaches that fall beyond the scope of this project. To isolate a few notable examples, this project neither offers a transnational comparison of LEGO products and advertising[31] nor conducts any substantive sociological or ethnographic research on actual LEGO users.[32] This is due more to expediency than desire, as to

[30] While these LEGO toys are compatible with the stud-and-tube brick system, they all rely heavily on alternative modes of play that push LEGO beyond the brick-based toy tradition. Featuring gears, motors, beams, and liftarms, Technic is more reminiscent of engineering toys like Meccano and Erector Sets than architectural toys. Building on this system, Mindstorms adds a computer module that allows players to animate Technic robots with LEGO-like block-based programming. Bionicle extended the Technic system in a very different direction, adding a fantasy story and redesigning the system for constructing mecha-like buildable action figures. Technic, Mindstorms, and Bionicle are all often written all in capital letters, but I reserve this notation only for the main brand name.

[31] Following the principle of localization, LEGO often targets particular products and advertising to specific regions based on national or linguistic affiliations. In the interest of maintaining focus, this project looks exclusively at English-language LEGO media targeted to a predominantly North American audience.

[32] To argue for the necessity of such work, Seth Giddings writes "To address the lived and moment-by-moment events of LEGO play requires ethnographic research with children and/or memory-work" (2014, p. 242). This is an extremely fertile avenue for future scholarship, but it falls outside the scope of this project which aims to deconstruct the medium and messages of LEGO play, an exploration of the systematic material and ideological design of LEGO texts. Indeed, I believe that these two approaches balance each

properly address these topics would require sacrificing a narrower focus on media-specific deconstructive analyses of core LEGO play, analyses which I hope will prove useful as further scholarship continues to develop the aforementioned approaches.

Play is serious business—in more than one sense in the case of commodified toys. And play matters all the more because its more serious significances are tied up in spontaneous, creative, joyful performances. Thus, despite the critical disposition[33] of this project, the following critiques rest in the persistent hope that the ideologies *at play* in LEGO are always also *in play*—that is, open to reinterpretation and transformation. Therefore, this deconstructive project disassembles particular ideological formations not to lay waste to all meaning-making in LEGO but rather to transform the conditions under which such meaning-making takes place, disempowering the ideological playscripts to instead empower critical, transformative, and generative play.

Seattle, USA

Jonathan Rey Lee

Works Cited

Althusser, Louis. 1971. *Lenin and philosophy and other essays*. Trans. Ben Brewster. New York: Monthly Review Press.

Bacharach, Sondra, and Roy T. Cook. 2017. Introduction: Play well, philosophize well! In *LEGO and philosophy*, ed. Roy T. Cook and Sondra Bacharach, 1–3. Hoboken: Wiley Blackwell.

Baichtal, John, and Joe Meno. 2011. *The cult of LEGO*. China: No Starch Press.

Bender, Jonathan. 2010. *LEGO: A love story*. Hoboken: Wiley.

other, as better understanding the medium and messages of LEGO will also inform how to organize ethnographic research into how these texts are received and reinterpreted by actual players.

[33]To some, the following analyses may seem to skew more toward exposing the ethical problems of LEGO's ideological constructions than recognizing its values or successes. This is largely accurate. Personally, I believe my responsibility as a media scholar is to be something of a resistant reader who raises ethical questions about corporate media. Rather than deny the ethical value of LEGO, I believe that such resistant readings help facilitate a practice of critical play that may further unlock the ethical potential that I believe LEGO to genuinely possess.

Brewer, John. 1980. Childhood revisited: The genesis of the modern toy. *History Today* 30: 32–39.

Day, Lori. 2014. The little girl from the 1981 LEGO ad is all grown up, and she's got something to say. *Women You Should Know*. https://womenyoushouldknow.net/little-girl-1981-lego-ad-grown-shes-got-something-say/. Accessed 1 April 2020.

Derrida, Jacques. 1998. *Monolingualism of the other; or, the prosthesis of origin*. Trans. Patrick Mensah. Stanford: Stanford University Press.

Dewar, Gwen. 2018. The benefits of toy blocks: The science of construction play. *Parenting Science*. https://www.parentingscience.com/toy-blocks.html. Accessed 17 January 2020.

Fanning, Colin. 2018. Building kids: LEGO and the commodification of creativity. In *Childhood by design*, ed. Megan Brandow-Faller, 89–105. New York: Bloomsbury Visual Arts.

Garlen, Jennifer C. 2014. Block party: A look at adult fans of LEGO. In *Fan CULTure*, ed. Kristin M. Barton and Jonathan Malcolm, 119–130. Jefferson: McFarland & Company.

Giddings, Seth. 2014. Bright bricks, dark play: On the impossibility of studying LEGO. In *LEGO studies*, ed. Mark J. P. Wolf, 241–267. New York: Routledge.

Gray, Jonathan. 2010. *Show sold separately: Promos, spoilers, and other media paratexts*. New York: New York University Press.

Hinck, Ashley. 2019. *Politics for the love of fandom: Fan-based citizenship in a digital world*. Baton Rouge: Louisiana State University Press.

Landay, Lori. 2014. Myth blocks: How LEGO transmedia configures and remixes mythic structures in the Ninjago and Chima themes. In *LEGO studies*, ed. Mark J. P. Wolf, 55–80. New York: Routledge.

Lauwaert, Maaike. 2009. *The place of play: Toys and digital cultures*. Amsterdam: Amsterdam University Press.

Lee, Jonathan Rey. 2014. The plastic art of LEGO: An essay into material culture. In *Design, mediation, and the posthuman*, ed. Dennis M. Weiss, Amy D. Propen and Colby Emmerson Reid, 95–112. Lanham: Lexington Books.

Lee, Jonathan Rey. 2019. Master building and creative vision in *The LEGO Movie*. In *Cultural studies of LEGO*, ed. Rebecca C. Hains and Sharon R. Mazzarella, 149–173. Cham: Palgrave Macmillan.

The LEGO Group. 2012. *The LEGO® story*. YouTube. https://www.youtube.com/watch?v=NdDU_BBJW9Y. Accessed 24 April 2020.

The LEGO Group. 2014. *A short presentation*. LEGO.com. https://www.lego.com/r/aboutus/-/media/about%20us/media%20assets%20library/company%20profiles/the_lego_group_a%20short%20presentation_2014_english_ed2.pdf. Accessed 7 October 2014.

The LEGO movie [Film]. 2014. Directed by Phil Lord and Christopher Miller, Warner Bros.

Lipkowitz, Daniel. 2012. *The LEGO book*. 2nd ed. London: DK.

McGraw, John J. et al. 2014. Culture's building blocks: Investigating cultural evolution in a LEGO construction task. *Frontiers in Psychology* 5: 1–12.

Moreau, C. Page, and Marit Gundersen Engeset. 2016. The downstream consequences of problem-solving mindsets: How playing with LEGO influences creativity. *Journal of Marketing Research* LIII: 18–30.

Prout, Alan, and Allison James. 2015. A new paradigm for the sociology of childhood?: Provenance, promise and problems. In *Constructing and reconstructing childhood*, ed. Allison James and Alan Prout, 6–28. New York: Routledge.

Robertson, David C. 2013. *Brick by brick: How LEGO rewrote the rules of innovation and conquered the global toy industry*. New York: Crown Business.

Sarkeesian, Anita. 2012. *LEGO friends—LEGO & gender part 1*. YouTube. https://www.youtube.com/watch?v=CrmRxGLn0Bk. Accessed 24 April 2020.

Sawaya, Nathan. 2014. LEGO: The imperfect art tool. In *LEGO studies*, ed. Mark J. P. Wolf, 206–215. New York: Routledge.

Acknowledgments

I cannot say whether I have whiled away more of my life playing with LEGO or writing about it. What I can say is that in both cases, countless quiet hours of making would undoubtedly have been impossible without incredible support. The freedom to undertake creative endeavors rests on material conditions that should never be taken for granted. So, I first want to acknowledge that my work and play alike rest on privileges I have never nor could ever truly earn.

Writing, like LEGO play, is a potentially lonely endeavor best conducted in community. As a young literature scholar with no training in analyzing play or media, I felt completely overwhelmed when this supposedly trivial side project began to snowball into a massive interdisciplinary undertaking. So, this project would have undoubtedly fizzled out without the grounding and guiding wisdom of many generous people.

I want to especially thank Alex Anderson, Meredith Bak, Heidi Brevik-Zender, Sabine Doran, Christine Harold, Dan Hassler-Forest, Steve Groening, Cameron Lee, Regina Yung Lee, Larin McLaughlin, LeiLani Nishime, the contributors who maintain the websites Brickset, Bricklink, and Brickipedia, the University of Washington Department of Communication, and the members of the 2019 SCMS Seminar on Toys and Tabletop Games. I also want to thank the talented professional and fan artists who contributed images to this project: Christian Bök, Mike Doyle, Olafur Eliasson, Paul Hetherington, Malin Kylinger, Aaron Legg, Steve Price, and Jeroen van den Bos and Davy Landman.

And most of all, I want to pour out my heartfelt thanks to my family, friends, and community group for such unflagging material, emotional, and spiritual support. To be able to write from a space of thriving is truly a great blessing. Finally, I want to dedicate this book to Soraya and all the children of this next generation. I hope when you play with your LEGO, you are inspired to build a kinder world than the one you are being born into.

Contents

List of Figures

CHAPTER 1

Theorizing LEGO Bricolage: Medium, Message, Method

Paradoxically intertwining the freedom of a reconfigurable medium with the scripted messages of a socializing toy, LEGO is a meaning-making system whose "possibilities always remain limited by the particular history of each piece and by those of its features which are already determined by the use for which it was originally intended" (Lévi-Strauss 1966, p. 19). This quote perfectly describes LEGO, despite actually referring to the practice of *bricolage* from which this chapter derives its theory and methods for deconstructing the medium and messages of LEGO play as a playful practice of meaningfully mixing and remixing extant meanings.

Take, for instance, the LEGO Parisian Restaurant (Fig. 1.1a), a case study in mixed messages. Consider the oddity in how—at once architectural and theatrical, static and dynamic—this set advertises itself both as a display model and as an interactive playset. Or how this visually busy and spatially cramped model serves as a metonym for the charmingly contradictory leisurely bustle of Parisian culture. Or how this highly localized Parisian locale literally fits into (note the connections on the bottom-right side) a modular building series of other structures that are almost entirely non-localizable. Or how regular and rigid plastic elements are cleverly used to create an aura of artisanal irregularities (especially note how the chimney disrupts its regular brick patterning by strategically placing differently colored bricks and raised elements).

Consider, moreover, the white LEGO croissants (Fig. 1.1b) deployed as architectural features on the restaurant's cornices. Here, the croissant is

J. R. Lee, *Deconstructing LEGO*,
https://doi.org/10.1007/978-3-030-53665-7_1

Fig. 1.1 **a** Product packaging for the 2014 Parisian Restaurant (Set #10243), part of a series of modular buildings within the LEGO Creator product line. Note the two pin connectors along the bottom right edge that allow this set to connect to other modular buildings (indicated by circles). Contextualizing this design is a blue-toned photographic background and multiple smaller images that show different angles. Of these, the lower-left also displays dimensions as an explicit reminder that this set is designed for display within the modular building series. **b** A magnified view showing the four white LEGO croissants (Item #33125) that flank the seashells on the cornices (indicated by circles)

essentially abstract, an atom of shape and color somewhat like a single dot in a pointillist sketch. At the same time, it is meaningful that this element is a *croissant* and not, say, a LEGO sausage (which would be materially interchangeable). Because LEGO elements are atoms of meaning that remain visibly distinct even as they are incorporated into constructions, the croissant features as a visual pun on a Parisian restaurant—both making and made of French fare. Reminiscent of wordplay based on the disconnect between sounds and meanings of words, this brick-play relies on the interplay of an element's form and content. Used in this way, the croissant becomes indicative of the playful spirit through which LEGO embraces its many underlying paradoxes and contradictions, a spirit that belies the serious ideological character of its play philosophy.

More than whimsical details, the mixed messages that run throughout LEGO design history also ideologically construct LEGO play through embedded *playscripts* that construct and instruct the playing subject in potentially problematic ways. This project deconstructs the ideological formation of five forms of LEGO play—*construction play*, *dramatic play*, *digital play*, *transmedia play*, and *attachment play*. To facilitate this, this chapter advances some theoretical and methodological foundations for deconstructing LEGO derived from the phenomenon of bricolage. Deconstructing LEGO requires more than 'reading' LEGO meanings because LEGO does not merely 'send' messages, mixed or otherwise. A toy medium cannot passively transmit fixed content as it is sometimes imagined mass communication media do (although this is too simple even for such media). Instead, a toy medium offers uniquely tangible invitation to play with and play out its embedded meanings.

The croissant may seem to send a clear message because it is visually embedded in another visual medium—product packaging—that provides a built-in interpretive context for reading its meaning. The croissant is materially 'transmitted,' however, as a loose physical piece that can circulate through countless actual and potential processes of meaning-making as it is alternatively collected, organized, assembled, disassembled, and toyed with. LEGO, that is, at once provides fixed messages and material building blocks designed for playfully remixing those messages (or the elements of those messages). In other words, this project deconstructs not only the mixed messages themselves but also the systematic way in which this interactive, participatory medium mediates its own construction and reconstruction. Rather than transmitted content to be passively received, the messages of LEGO are implied ideological frameworks that inflect the performance of creative play.

The mixing of messages in LEGO is rarely incidental and never insignificant, as these mixed messages speak to forces that inflect the creative possibility spaces that define LEGO play, suggesting a constitutive tension between freedom and constraint. Thus, while one LEGO Foundation research report asks, "How do we innovate if we also conform to socially shared and inherited conventions?" (Alsdorf and Gravel 2018, p. 58), another concludes that precisely by establishing such social conventions, "the LEGO Group can play an active role in … better highlighting the value of free play, tinkering, exploration and discovery," as "The growing social dimension of play also calls for more action to connect phases of the play and creation process socially, supported by platforms and ecosystems where all members can contribute and find value in participating" (Gauntlett et al. 2010, p. 5). As imagined by the LEGO Foundation, creative play demands a supportive context that socially organizes and thereby ideologically inflects creative development.

Just as the LEGO System facilitates creative possibilities tied to material constraints, this LEGO culture attempts to facilitate creative play precisely by constraining it. Here LEGO draws from psychological research[1] that argues: "Not all play is created equal. Guided play in particular, where an adult scaffolds a situation toward a specific learning goal, may be especially helpful at maximizing engagement, particularly for younger children who are more susceptible to distraction" (Zosh et al. 2018, p. 5). Guided, scaffolded, and scripted, LEGO play offers possibilities for mixing and remixing messages within an implicit structuring context that strongly suggests what LEGO play can and should mean.

Unraveling this system of mixing and remixing messages requires some critical foundations—even a deconstructive analysis plays off the distinctive ways LEGO makes meaning. Outlining even a few basic elements of these foundations is a substantial task for a toy medium and media toy that neither fits neatly into existing critical paradigms nor has quite yet established a large critical discourse of its own. To this end, this chapter sketches a few relevant theoretical and methodological threads that might inform such an analysis.

[1] Although this article is not published by the LEGO Foundation, one of its co-authors was a member of the Foundation when this article was published.

The first section frames the study of LEGO within some key ideas from Media Studies, posing *bricolage* as a central metaphor for the media-specificity of LEGO. The following section theorizes the ***medium*** of LEGO by unpacking how the material design of the LEGO System facilitates a distinctive process of bricolage that synthesizes the material paradigms of atomism and plasticity. The third section explores the *messages* in and around the medium by delving further into the symbolic economy of LEGO to show how its meaning-making elements engage larger cultural ideologies. And finally, the fourth section lays out some methods for analyzing the elements, sets, and paratexts that make up the LEGO media landscape.

Mediating Bricolage

Although media scholars often emphasize how our contemporary media culture increasingly encourages media consumers to also become producers, there is often a clear material separation between media consumption and production. Most media texts available for mass consumption—books, films, artworks—have fixed content that cannot be easily edited. Conversely, most tools for producing these media—pen and paper, video cameras, art supplies—contain no consumable content whatsoever. This kind of material separation of production and consumption is unimaginable for LEGO, whose products are at once media texts and toolkits for producing media texts. Thus, one could picture the croissants in the Parisian Restaurant set as if each brush stroke in "Starry Night" could be peeled off and stuck onto other paintings. Any media theory of LEGO will, therefore, have to account for how this medium intertwines production and consumption through the constitutive mixing and remixing of extant meanings. To express this, the following media theory of LEGO is built upon the concept of *bricolage*, a form of creative practice based on the *creative reassembly of already-significant elements*.

While any metaphor distorts even while it lends insight, the one I have found most helpful for conceptualizing LEGO is *bricolage*, a French term for the humble yet creative practice of the *bricoleur* or "handyman." The closest parallel to this practice in English is *tinkering*, which derives from the similar profession of the *tinker* (itinerant repairman) in the Middle Ages. Tinkering is a compelling image for modern and especially postmodern creative subjects who often create with and within an expanding collection of scraps scrounged from consumer society. Among other things, tinkering is material, quotidian, tactical, creative, transformative, resourceful, recuperative, constructive, and playful.

Reflecting these traits, bricolage commonly names artistic styles and techniques based on *assemblage*, such as sculptures pieced together from found objects and jumbled-together architectural styles. Similarly, without coalescing into any single consistent use, the metaphor of bricolage has been appropriated by quite varied theorists[2] to refer to cultural or media practices that appropriate and remix items in surprising, creative, or non-linear ways—for the sake of clarity, I synthesize these various usages into a single theoretical formulation: the *creative reassembly of already-significant elements*. This formulation is particularly well-suited to analyzing LEGO, a system for mixing and remixing its elements through playful tinkering. Thus, whereas any medium can be marshaled for bricolage, LEGO is a medium *of* bricolage—material play in LEGO is entirely oriented around bricolage as creative practice.

To adapt this concept of bricolage to the study of LEGO, it will be helpful to first situate this theory within several approaches to media scholarship that themselves need to be creatively reassembled to account for this medium of mixed messages. The first approach, dubbed *media-specific analysis* by N. Katherine Hayles (2004), considers how the distinctive characteristics of any medium inflects its messages. This more localized approach encourages interpreting media texts to account for the specific materiality of the textual object. This means not only interpreting the visual effect of the Parisian Restaurant as if it were any other miniature object but also interpreting how this set plays off its uniquely LEGO construction. For instance, the primary structure of this set builds on traditional LEGO architecture with its brickwork and specialized architectural elements, such as windows, fences, and flowers. At the same time, this set demonstrates "expert" building by liberally using SNOT (Studs Not On Top) techniques and cleverly repurposing unexpected elements, such as the aforementioned croissants and planters filled with green "spike" swords (Item #64727). Developing such media-specific analysis will prevent the study of LEGO from being beholden to interpretive lenses developed for quite different media and, instead, equip a study of the distinctive ways this toy medium and media toy makes meaning.

[2] Notable examples include Claude Lévi-Strauss' sociological work on comparative religions, Dick Hebdige's work on punk subculture in *Subculture: The Meaning of Style*, and Sherry Turkle's work on programming education in *Life on the Screen*.

While *media-specific analysis* methodologically informs how this deconstructive project 'reads' LEGO texts, a second approach provides a wider lens on how LEGO participates in the ongoing formation of larger *media cultures*. Inspired by[3] Marshall McLuhan's (1964/2005) historic maxim "the medium is the message," this approach is less concerned with analyzing the specific messages being conveyed in a medium and more concerned with tracing the wider cultural resonances of the medium itself. A medium, after all, transcends its particular messages to constitute a possibility space for meaning-making. And this potentiality is itself culturally impactful. Thus, just as McLuhan argued that no particular television show transformed culture to the extent that the very presence of the television in the home did, one could argue that the primary cultural impact of LEGO is how the LEGO System as a whole has transformed play. From this perspective, the media-specificity of LEGO not only determines how particular LEGO texts make meaning but rather lends shape and structure to an entire culture of play. To map this cultural geography of play, this project traces five culturally-impactful modes of play that the medium is designed to facilitate, ways the LEGO system scripts playful performances toward playing out specific cultural ideologies.

Finally, a third approach bridges these approaches by exploring how LEGO participates in *cross-media* or *transmedia ecologies* with other non-LEGO media. Neither the media-specificity nor the cultural formation of LEGO exist in a vacuum—instead, LEGO is continually formed through dialogic relationships with other media, media texts, and media paratexts. Indeed, the LEGO brand has extended well beyond the toy into surprisingly diverse media including film (see Chapter 6), television (see Chapter 5), videogames (see Chapter 4), comics, theme parks, web content, augmented reality apps, merchandise, and more. Indeed,

[3] Although this project draws on McLuhan's general approach, which I am calling a *media cultures* approach, it does not hold exactly to his methodology or terminology. First, while McLuhan has sometimes been criticized as being too technologically determinist, this project does not consider LEGO as straightforwardly causing cultural impacts but rather as being materially implicated in various cultural trends that are influenced by a multiplicity of factors. Second, this project explores how the medium can function as a kind of message while still performing more traditional readings of the specific ideological messages conveyed by LEGO texts. In other words, it both treats the medium as the message and its messages as the message. Finally, although McLuhan uses the language of *media ecologies* to describe his approach, I use the term in a much more general way to describe a broader trend of ecological thinking that has resonated throughout a variety of media scholarship.

one of the more interesting aspects of the LEGO media ecology is the unusual way it blends toylike media participation with traditional media consumption.

LEGO, therefore, exemplifies what Henry Jenkins (2013) has called "participatory media culture," a culture of media consumption in which consumers have increased ability to directly contribute their own media content, often engaging in the *tactics* that Michel de Certeau (1980/1988) describes as micro-resistances to the *strategies* of dominant culture. Whereas productions like fan fiction, fan vidding, and cosplay maintain a clear separation between the canonical and fan-produced texts, the simple physical act of snapping a single LEGO piece into place is a moment of textual production that necessarily blends corporate and fan authorship. As LEGO is participatory in an inescapably tangible way, LEGO consumption cannot be conceptualized without production.

Although media scholarship often emphasizes one of these approaches above the others, a theory of LEGO bricolage draws out the generative complementarity of all three approaches. Bricolage anchors LEGO's media-specificity, characterizes its media cultures, and structures its transmedia engagements. The creative reassembly of already-significant elements resonates throughout the medium and messages of LEGO.

As a medium of *material bricolage*, LEGO play literally enacts this creative reassembly in ways that closely resemble the activity of the traditional bricoleur. And as a medium of *symbolic bricolage*, LEGO play evokes and reassembles elements of cultural ideologies in ways that resembled how the traditional bricoleur appropriates elements that are "pre-constrained," as Lévi-Strauss explains, "like the constitutive units of myth" (p. 19). To ground this project, therefore, the following two sections draw upon theories of bricolage to expand on the material and symbolic bricolage that respectively characterize the medium and messages of LEGO play.

Medium: Material Bricolage

Media are material—to mediate is to materially record, transmit, or otherwise express. Media-specificity is grounded first and foremost in the material affordances and constraints that define media as meaning-making possibility spaces. Thus, the LEGO medium is first and foremost a system of material bricolage. Far from being abstract, however, the materiality of LEGO is precisely where its ideological foundations lie. After all, while materiality and ideology are intertwined in any designed object, this is doubly true for a patented construction system with a strongly defined brand identity like LEGO. Thus, this section theorizes LEGO's material mediations to provide a framework for deconstructing the multi-faceted ideological construction of LEGO, a systematic toy medium based on mixing and remixing atomistic yet plastic elements into meaningful assemblages.

While all media are material, few emphasize the intimate materiality of touch like LEGO—even sculpture typically invites visual rather than tactile consumption. As I have argued (2014), by toying with atomism and plasticity, LEGO makes graspable the fundamental *int(e)ractability* of matter itself, how the very qualities that make matter materially intractable define its potential to be materially interact-able. Similarly, the media-specificity of LEGO relies on the material affordances that make the medium interact-able resting in the very material constraints that make the medium intractable. LEGO bricolage invites players to directly touch and manipulate this contested possibility space, the systematically int(e)ractable push-and-pull between material affordances and constraints.

This push-and-pull is incredibly visceral and generative—so much so the one LEGO Foundation research report argues that 'the world pushing back on us' is central to how humans explore the material world:

> And, as we work with objects in the world, they continually "push back on us." We try and make a piece of wood bend and it won't. We try and make a tower of blocks that stand, but it falls. These are examples of objects in the world pushing back on us, and they and objects and how they work. This pushback creates opportunities for us to learn because these are moments where things do not behave as we expect, and that tension drives us to make sense of things. (Alsdorf and Gravel 2018, p. 12)

As a toy, LEGO offers children tangible ways to explore this tension and "make sense of things." Unlike a tower of blocks tumbling over, however, LEGO does not offer exploration of general physical properties like gravity. Instead, LEGO explores its internal principles of material int(e)ractability, including its engineered system for holding such towers together. LEGO is carefully designed to help players 'make sense' of its own construction. Material bricolage is not only the means of LEGO play—it is also its end, a characteristic goal of LEGO play being the pleasurable exploration of the material constraints and affordances of its play system. It matters, therefore, not only *that* but also *how* LEGO constitutes a medium of material bricolage.

LEGO is a medium of bricolage first and foremost because, as a construction toy, it is designed for material assembly and reassembly.[4] The meaningfulness of any construction toy lies in the combinatory potential of elements that exponentially multiply possibilities when added to the system. In contrast to some other construction toys, however, the media-specific interconnectivity of the LEGO System multiplies these possibilities at a remarkable rate. Just six LEGO bricks of the same color can be combined in 915,103,765 distinct ways, an astronomical yet finite number that affords a much more varied possibility space than the technically infinite possibilities of six traditional building blocks.[5] In other words, a LEGO element contributes two types of meaningfulness: the formal qualities of the material element itself (shape, color, texture, etc.) and the connective potential defined by its patented stud-and-tube design (Fig. 1.2). LEGO, in other words, simultaneously offers material elements for inclusion in the act of bricolage and inherent physical principles that define the rules for their assemblage.

In this way, LEGO is both like and unlike other construction toys. Although all construction toys are media of bricolage in a similarly systematic way, the specific proprietary design of each construction toy system provides a strikingly distinct range of material possibilities. The specific affordances and constraints that define LEGO differ noticeably from the nuts-and-bolts system of Meccano, the spool-and-rod system of Tinkertoys, or the notch system of Lincoln Logs.

[4] As Tamar Zinguer (2015) points out, "if the term 'bricolage' were applied to the activity of building with construction toys, it would imply toying with the 'odds and ends' of architecture, playing with the 'remains and debris' of architectural history. Overall, then, the bricoleur is an apt description of the player—the child—on whom the detritus of architecture has been bestowed" (205). This speaks both to the symbolic bricolage that ties construction toys to architectural history and to the material bricolage that defines the systematic interconnectivity underlying construction toy play.

[5] Traditional blocks that lack connective potential offer a theoretically infinite number of combinations, but this infinity derives from the inexhaustible preponderance of imperceptible variations in placement. These variations do not always lead to meaningful differences. So, although the constraints of the connective system mean that two LEGO bricks can be combined in a finite number of ways, each different placement is more clearly distinct and potentially meaningful. For more information on this statistic and how it was calculated, see *A LEGO Brickumentary* (2014).

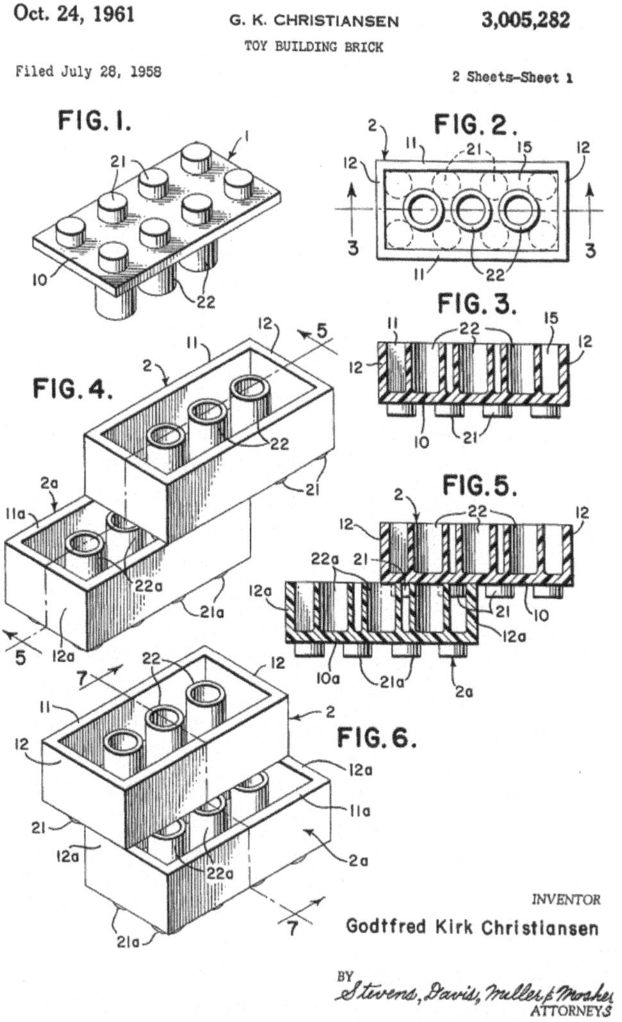

Fig. 1.2 The material characteristics of the LEGO brick are determined by its patented (see 1958 patent above) *stud-and-tube* design, which modifies Hilary Page's earlier Kiddicraft Toy Building Blocks to provide better *clutch power* (strong connections that are nevertheless easy to detach). The *studs* (#21)—which each bear the LEGO name (not pictured here)—snap into the gaps between the sides of the LEGO bricks and the *tubes* (#22). This connection is made possible by the slight elasticity of the plastic Acrylonitrile butadiene styrene (ABS)

The LEGO System typically needs no introduction, as it is immediately recognizable to a significant percentage of the globe. Yet, a very brief sketch of its most defining features includes small, blocky, primary-color shapes with raised studs and inset tubes, small, blocky, cartoonish poseable minifigure characters and their accessories, a material pliability that makes the hard, sharp pieces 'snap' together, a horizontal orientation that spaces construction along a two-dimensional Cartesian grid, a vertical orientation in which construction primarily takes place by stacking, and so on. Theorizing the unifying logic of these traits, I characterize the distinctive material int(e)ractability of LEGO as a paradoxical interplay of *atomism* and *plasticity* (Lee 2014). As is typical of the mixed messages of LEGO, atomism and plasticity are simultaneously opposed and intertwined to produce a playfully hybrid material experience.

The atomism of LEGO rests in its systematic construction as the assembly of fundamental particles which become significant within a larger system of connections. This atomism is an invitation to explore the constructedness of a toy world by deconstructing and reconstructing its constituent elements. This reflects the compositional logic of traditional bricolage in which the meaning of the whole is visibly made up of component parts that retain some distinctiveness rather than disappearing into the whole. Moreover, whereas most artistic bricolage (sculptural bricolage, mosaic, pointillist art) assemble elements that can still be broken down further, atomistic LEGO elements are indivisible. As assemblages of these solid shapes persistently reveal their seams, the atomism of LEGO simultaneously offers its particular brand of bricolage while also serving as a material metaphor for bricolage as a whole.

This tangible, atomistic sensation of building a singular object from a multiplicity of other singular objects paradoxically also contributes to LEGO's plasticity. That just six bricks can be combined in far more ways than a person could assemble in a lifetime makes LEGO feel surprisingly fluid and malleable. Whereas the girders and rivets of Meccano suggest an industrial connectivity and the spools and rods of Tinkertoy suggest a molecular connectivity (they resemble chemical models), LEGO offers a more fluid, plastic feel through the tightness with which the pieces fit together as contiguous parts of a single form. At smaller scales and with vast connective possibilities, LEGO elements strike a distinctive balance that plays a high degree of visible and tangible atomism against a high degree of visible and tangible fluidity.

Together, the intertwined atomism and plasticity constitute LEGO as a medium of material bricolage in which the creative reassembly of already-significant elements takes the form of a fluid, plastic reassembly of intractable atomistic elements. As early childhood object play often derives pleasure from tangibly probing the fundamental int(e)ractability of matter—the push and pull of objects responding to and resisting the touch[6]—LEGO play derives its pleasure from conceptually and tangibly probing its own int(e)ractability as a construction toy system. With this self-reflexive orientation, LEGO is a medium whose message is always in part the medium itself. And as this medium rests on a constitutive material paradox, it is perhaps unsurprising that LEGO is also a medium of mixed messages, as the following section on the symbolic economy of LEGO explores.

Messages: The Symbolic Economy of LEGO

Ideological as well as material, int(e)ractability also underlies the symbolic economy of LEGO bricolage. Just as the material intractability of the LEGO System creates interact-able potential for material construction, the symbolic intractability of LEGO's embedded signifiers creates interact-able potential for symbolic meaning-making. Because the LEGO medium is proprietary to an extent that most mass or artistic media are not—after all, a specific film camera or paint formulation can be patented, but the media of film and painting themselves cannot—LEGO is a medium that is unusually saturated with corporately-authored messages.

Consequently, it is impossible to disentangle the reassembly of extant material LEGO elements from the reassembly of LEGO-branded social messages. LEGO is a medium with messages already included. Not at all a medium of abstract form like painting, LEGO is a system for reassembling already-significant corporately-designed symbolic elements, like written language might be if writers had to individually purchase words from a single supplier. Thus, material LEGO elements function as individual semantic units within a construction system that functions like a material

[6] In "Beyond the Pleasure Principle" (1920/1989), Sigmund Freud explores a game of *fort/da* in which a child pushes away an object for his mother to retrieve. The psychoanalytic argument Freud offers to explain this behavior is well beyond the scope of this project, but the example does illustrate how tangible push-and-pull play influenced the founding of modern psychology.

grammar. To disentangle this symbolic economy, this section theorizes how the messages of LEGO offer *playscripts* that ideologically condition LEGO play.

While material LEGO play seems an obvious candidate for a theory of bricolage, most scholarship that explicitly applies theories of bricolage to LEGO focuses primarily on LEGO branding. In arguably the first substantive piece of LEGO scholarship, Maaike Lauwaert (2009) draws upon Michel de Certeau's theories (including bricolage) to explore the dynamic relationship between corporate authorship and fan participation in LEGO. More recently, Dalia Grobovaite (2017) reads LEGO bricolage as an expression of how the global culture industry commodifies the very ideas it advertises as resisting commodification (such as individualism, creativity, etc.), thereby undermining such resistance. Peter Gregg traces a similar way LEGO approaches media franchising as a form of "brick-olage" that allows LEGO "to shape and be shaped by other brand styles, and so the toys epitomize a kind of postmodern play: a mosaic of brands and styles that, when converted to LEGOs, become subordinate to the rules of the toy" (Gregg 2018, p. 41). These examples productively demonstrate why branded LEGO cannot be reduced to its abstract material form but should instead be analyzed as embedded in a corporately-inflected media ecology.

These analyses speak to the cultural significance of toys as mediating cultural ideologies through children's play. Toys are not merely props—passive objects entirely determined by the imagination of the player. Toys *script* play by providing normative guidelines that, as Lauwaert claims, "make certain things possible and not others. Scripts are embedded in an artifact during the design process of this artifact" (2009, p. 13). That is, the designed materiality of the thing is also a symbolic economy that conditions playful performances. Indeed, psychological research loosely connected to the LEGO Foundation speaks to the importance of scripts, arguing that "Guided play, in particular, can be described as 'constrained tinkering' where, within a bounded exploration space, children have the freedom to test out different hypotheses" (Zosh et al. 2018, p. 7). This reveals the underlying rationale of playscripts, namely that constraints facilitate creativity especially in the case of bricolage or tinkering.

Scripts are both invitations to enter into particular ideologies and implied directions for how to perform such ideologies. At the same time, scripts admit improvisation as they are creatively interpreted and implemented by individual actors. In her insightful theorization of toys as *scriptive things*, Robin Bernstein explains that "*script* denotes not a rigid dictation of performed action but, rather, a necessary openness to resistance, interpretation, and improvisation" (2009, p. 68). In this paradoxical openness to resistance, one might say that scripts invite improvisation not by abolishing structure but by providing an inherently permutable structure. The normativity of scripting—suggestion rather than compulsion—thus closely resembles the implicit power of ideology. Scripted toys, in other words, are ideological outpourings of material culture.

To play with a proprietary, branded, symbolic toy like LEGO is to simultaneously consume and produce ideological content. Whereas many media exhibit a disconnect between consumption and production as media participants *alternately* consume and produce media content, LEGO players *simultaneously* consume and produce LEGO every time they snap a piece into place. Such play is necessarily an interplay between the significances baked into the toy medium and the creative reassembly of such significances in symbolic bricolage. The inseparability between the material potentiality of the toy medium and the symbolic economy of the media toy, therefore, constitutes the primary ethical problematic of LEGO, a medium designed to make meaning from (and according to) its many messages.

To help visualize how LEGO players/bricoleurs act as "(co-)designers and (co-)producers" (p. 18), Lauwaert proposes the metaphor of a LEGO *geography of play*, a shared terrain that underlies individual acts of creativity. Rather than a neutral, natural landscape, however, this geography is a contested space whose central tension Lauwaert describes as between *core* and *peripheral* play, a distinction that helps clarify LEGO's synthesis of medium and message. She characterizes core play as "constituted of facilitated play practices" which "enable and assist but also to promote, encourage and catalyze" certain forms of play (p. 12). In contrast, the periphery of play "is comprised of divergent practices of play that deviate from the discourse on a toy or game or that use the design of a toy or game in unexpected ways" (p. 13). It is noteworthy that the language of 'core' and 'periphery' does not imply an absolute distinction like 'inside' and 'outside.' Instead, this language implies a continuous progression between different degrees of proximity, aptly reflecting how the core is immanent in the periphery (since every peripheral use is a dialogue with the embedded design of the elements) and the periphery is imminent in the core (since every core use requires active user participation).

Core and peripheral play exist in perpetual tension, as LEGO play is conditioned by simultaneous yet contradictory imperatives toward creative and scripted play. This is well illustrated by the nesting of LEGO fan videos within *The LEGO Movie*. In one scene (Fig. 1.3), Lord Business complains that the citizens' creativity interjects chaos into his otherwise ordered world while the four background screens each show a stop-motion video submitted by a LEGO fan. Although this scene is a paradigm case of a large media corporation rewarding its fans with recognition while simultaneously capitalizing on their free production, it differs from most such appropriations by being situated within a framing narrative that is itself an allegory for the relationship between fans and company. As I argue elsewhere (2019), the film proclaims the value of everyday creativity, normalizing certain kinds of creative interventions into its designed system and thereby positioning such peripheral play as an extension of core play.

As a medium of both material and symbolic bricolage, LEGO advertises its core play as systematically inspiring creative, peripheral, playful permutations of its system. LEGO is at once a tool for mixing and remixing messages and a contested ideological construction shaped by

Fig. 1.3 Screenshot of *The LEGO Movie* depicting Emmet looking at four screens that show fan-submitted stop-motion videos (also known as Brickfilms; see Chapter 5 Post-script). As the frame is centered on the out-of-focus back of Emmet's head, the audience is invited to look past Emmet and instead share his vision of the four screens, which remain unobstructed

mixed and remixed messages. The material symbolic economy of LEGO bricolage, that is, unfolds as a complex interplay of forces that collapses distinctions between consumption and production, core and periphery.

Method: Deconstructing LEGO Texts

The ideological formations at play in this symbolic economy certainly call for deconstruction (see Preface). At the same time, how LEGO's peculiar blend of material and symbolic bricolage plays out in its branded products need reframed deconstructive methodologies. To this end, this section offers short methodological primers for 'reading' scripted LEGO products according to three foundational facets of LEGO's symbolic economy: *elements*, *sets*, and *paratexts*. Together, these facets constitute the multi-layered space of material and symbolic int(e)ractability through which LEGO mediates and thereby scripts its play.

Elements

The LEGO medium consists of a system of interconnecting *elements* ranging from basic bricks to specialized elements like croissants that can be (re)configured in different ways. The expressive potential of the medium directly results from the affordances and constraints of these elements, each of which contributes unique material and symbolic potential.[7] LEGO elements are form *as* content.

Materially, these elements function like Democritean atoms: rigid particles that materially manifest their own connective possibilities—they are objects of bricolage that materially structure the rules of their own construction (Lee 2014). Symbolically, these elements are more or less representational signifiers or icons, fragments of meaning that circulate within the symbolic economy of LEGO play—the material substrate bears embedded semiotic subtexts. Blending form and content across their material and symbolic layers of meaning, these int(e)ractable elements

[7] There is a continuum with respect to how strongly each element emphasizes its material or symbolic contributions. A classic 2 × 4 brick contributes quite a bit of material connectivity (8 studs on top and 8 spaces for studs on the bottom offers 16 places where this brick can connect) with relatively little symbolic content—but not none, as even the basic brick represents an architectural element rather than an abstract block. The croissant offers less connective potential (two endpoints which snap into LEGO hands or clips and one place for a stud to connect) and rather strong symbolic content.

constitute an object language, particles of meaning that contain their own grammar.

LEGO play involves intimately and tangibly reassembling these literal bits of material and symbolic meaning into new configurations. Designed for bricolage, these elements have relatively little play potential individually. Instead, they gain significance primarily when mobilized within larger assemblages of LEGO elements. Consequently, these elements are designed to circulate within the sets and paratexts that enmesh LEGO play in broader systems of social significance. To 'read' a LEGO element, therefore, is not only to read a singularly contentful form but also to read the vast possibility space of that form's implied circulation.

Sets

The defining LEGO products, LEGO *sets* consist of groups of elements packaged with pictorial instruction manuals for constructing specific structures (typically buildings, scenes, or vehicles). While it is possible to purchase individual elements directly from the company or aftermarket retailers, most LEGO circulates through these sets, which therefore constitute the central authorized LEGO texts.

Aided by the immediate paratexts of building instructions and product packaging, sets guide the player through first through a linear process of construction and second through forms of play implied by the set design. The set, therefore, requires a special kind of close reading: as a built environment whose design *scripts* (as well as enables) how it is to be played with. As a toy, sets function both as objects of play and as messages *about* play, which can be read through an object analysis akin to those proposed by classical semiotics, material culture studies, toy studies, thing theory, and so on.

At the same time, unlike most other scripted toys, the semiotics of the LEGO set also involves the process of its construction. LEGO, that is, makes meaning through what Ian Bogost (2007) calls *procedural rhetoric* (how procedures or algorithmic sets of rules can present messages and even advance arguments). In this way, sets imply an activity that invites process-based approaches to media-specific analysis akin to those in cultural studies, play studies, and game studies. To 'read' a LEGO set, therefore, entails both reading how the architecture of a constructed set scripts dramatic play (see Chapter 3) and reading how the process of constructing a set itself procedurally shapes how players engage the medium.

Paratexts

Literary theorist Gérard Genette (1997) coined the term *paratexts* to describe the myriad ways in which our understanding of print media is shaped by the supplementary texts that surround and contextualize it, such as the book cover, title, genre classifications, etc. Jonathan Gray (2010) extends this line of thought to contemporary media culture by analyzing how media texts are filtered through preconceptions shaped by advertising, critical reviews, hype, and other forms of second-degree encounters. Paratexts frame how to engage media, which I argue is especially significant for participatory, interactive, play media like LEGO whose significance is especially dependent on how they are engaged.

A book or film repackaged in a plain, unbranded cover is still largely the same text. By contrast, the elements of a LEGO set repackaged in a plain box are no longer remotely the same text. LEGO sets are organized by two constitutive *paratexts*—building instructions and product packaging—that script a single canonical construction from the practically infinite possibilities one box of elements enables. These essential paratexts translate LEGO elements into LEGO sets.

Further out, other paratexts circumscribe the culture of LEGO play and the formation of the LEGO brand identity. LEGO is thoroughly circumscribed by messages presented in retail stores, multimedia advertising campaigns, LEGO-sponsored clubs and events, parenting and educational resources, popular television and videogame franchises, and so on. Some of these paratexts, such as LEGO videogames (Chapters 4 and 5) and *The LEGO Movie* (Chapter 6), are well-developed texts in their own right. For these examples, the distinction between text and paratext quickly collapses because, as Gray argues, what makes one text the 'text' and another the 'paratext' is merely a matter of perspective.[8]

[8]This caveat, like Einstein's idea that in physics there is no privileged reference frame, does not mean that all texts are received as having equal weight. Stories often circulate in ways that privilege one text or medium over others. I argue that despite growing its multimedia production, LEGO still privileges its playsets as its dominant texts. Thus, for example, when the LEGO videogames depict adventures in famous story worlds (from *Star Wars*, *Harry Potter*, *Lord of the Rings*, etc.), I argue that the primary logic behind their world building is neither that of the source texts (film, literature) nor that of their current medium (videogames), but rather that of physical LEGO toys (see Chapter 5). That is, LEGO videogames exhibit a toy-centric design by presenting their physical spaces as to be constructed and, more commonly, deconstructed.

Whether in primary or supplementary roles, these paratexts advance messages about play that help shape the cultural context (and market) within which LEGO takes on meaning. To 'read' LEGO paratexts certainly requires media-specific analysis of its particular media form. After all, paratexts use their distinctive textuality to speak to LEGO toys. Yet, reading LEGO paratexts also requires deconstructing uniquely transmedial moments (see especially Chapters 4–6) in which LEGO reconstructs other media-specific forms according to a toy logic. Treating the ethos of the LEGO toy as a defining characteristic of the LEGO brand, LEGO develops unusual paratexts that emphasize aspects of their media-specificity to draw out the media-specificity of its central toy texts.

To summarize, LEGO play is always already structured by the intersecting facets—elements, playsets, paratexts—of its material and symbolic design. Yet, unlike the compositional logic commonly associated with LEGO, the relationship between these facets is fluid.[9] Interlacing symbolic significances into a material system that structures its own construction, LEGO infuses layers of meaningfulness into the meaning-making process itself.

To deconstruct LEGO is to analyze how the ideological construction of LEGO is woven through these three intertwined facets, each of which speaks to the material and symbolic int(e)ractability that constitutes the LEGO medium. Most media offer spaces of possibility for conveying an infinite array of messages. By contrast, LEGO offers spaces of possibility for playing with the actuality of messages already embedded in the medium. A medium of material and symbolic bricolage, LEGO promises playful encounters with the creative reassembly of already-significant elements.

In the spirit of playful bricolage, this project appropriates, adapts, and analogizes many critical discourses, both here and in the wide range of

[9] LEGO design is a kind of bricolage that engages all these facets. LEGO designs new elements to fit trends in playsets or marketing, new playsets that creatively reuse elements to expand marketing lines, and new paratexts to advertise their new elements and playsets. Similarly, the LEGO player will primarily encounter these facets as interrelated. The player who first encounters LEGO through the paratexts of advertisements or box art will simultaneously encounter the sets and elements displayed within those representations. And the player who first encounters LEGO jumbled together in a family member or friend's collection will encounter a mix of elements that bear visible traces of their function within sets and paratextual discourses.

cultural ideologies explored in the following chapters. Indeed, this piecemeal approach is part of the point. Although there is certainly some value to theories and schemas (especially for a topic that has as yet no established critical vocabulary), there is a certain aptness to studying play—especially the perpetually reconfiguring medium and messages of LEGO—performatively.

Rather than aiming to concretize a universal theory of LEGO, this project instead aims to catalyze critical interventions into a complex, compelling, and often problematic medium and its messages. Accordingly, the following chapters not only constitute such interventions themselves but also hope to incite further interventions. Deconstructing LEGO, that is, is more than merely articulating static or fixed messages—it is an active process of engaging the ongoing cultural formation of a paradoxical toy medium and media toy, an ethical provocation that—like LEGO—is predicated on connection.

Post-script 1—Christian Bök's Poetic (De)Composition

More self-reflexive than many media, LEGO advertises its nature as a medium of bricolage as a core aspect of LEGO play. At the same time, these consumer-facing invitations to play in particular ways do not always critically interrogate the complex dynamics of bricolage. Contemporary art sometimes performs a more deconstructive form of self-reflexivity, using the LEGO medium to critically interrogate itself. The best-known example of this is Nathan Sawaya's *Yellow*, a humanoid LEGO bust made of basic yellow bricks opening up its chest so a stream of bricks can pour out. As I have written elsewhere (2014), this sculpture illustrates the compositional relationship between the atomism and plasticity of LEGO to interrogate the compositionality of the human form. Applying a similarly deconstructive moment to a wider scope, Canadian poet and artist Christian Bök creates LEGO art that deconstructs the fundamental *compositionality* of LEGO, language, and the physical universe. Whereas Sawaya's art draws out the sculptural form of LEGO (see Preface), Bök's more experimental art leverages the paradoxical intersection of atomism and plasticity to raise questions about how assemblages construct meaning from their component parts. Thus, Bök uses artistic bricolage to interrogate the theoretical foundations of bricolage itself.

Bök poignantly unravels recombination as the essence of mediated meaning-making in two experimental mixed-media art pieces that pair LEGO with poetry built from anagrams. In *The Great Order of the Universe*, Bök pairs a matrix of hand-drawn sketches representing ways to connect LEGO bricks with a pair of quotes—one from Greek atomist Democritus and one from the original LEGO patent—that, as anagrams, consist of the same linguistic elements in different combinations. Drawing upon the eerie resonance between these two radically removed moments in history, Bök experiments with the idea that the 'great order of the universe' follows an essentially atomistic logic—that differences in being are reducible to differences in relation, combination, and assembly.

Similarly, in *Ten Maps of Sardonic Wit* (Fig. 1.4), Bök constructs a

Fig. 1.4 *Ten Maps of Sardonic Wit* by Christian Bök. Each page features a brick mosaic spelling out a line of poetry, all of which are anagrams of each other (Image courtesy of the Marianne Boesky Gallery with permission of the artist)

material book of poetry out of 'pages' built of black-and-white LEGO mosaics. Treating language as LEGO, the poem consists of ten lines which each provide one 'map of sardonic wit' by being an anagram of the poem's title. Connecting atomism with poetic composition, Bök slips between cosmic and linguistic imagery, writing that "atoms in space now drift / on a swift and epic storm" and "words spin a faint comet." Speculating that "some words in fact paint / two stars of an epic mind," Bök elides medium and message, explaining that his poem both employs and comments on the reassembleability of the universe, language, poetry, and LEGO:

> Each line is an anagram that exhaustively permutes the fixed array of letters in the title, recombining them into a coherent sequence of statements about the relationship between atoms and words. The poem suggests that just as permuted elements can create compounds, so also can permuted phonemes create syllables. The letters of the poem become the literary variants of subatomic particles, and the book itself embodies these molecular metaphors, insofar it too consists of discrete elements that can be dismantled and recombined to form a radically different structure. (2012, n.p.)

Mediating between atoms (or universe) and language, LEGO provides the medium for such experimental art because the media-specificity of LEGO contributes special significance to Bök's poetry. As a medium of bricolage, the possibility of constructing and reconstructing LEGO is, in a sense, inherently anagrammatic—an exercise in discovering new meaningful combinations of a finite number of fixed elements. Bök thus finds poetic potential in how the material grammar of LEGO bricolage resembles the linguistic bricolages of words comprising sentences and letters comprising words. For Bök, poetic *composition* (literally 'placing together') necessitates bricolage.

Probing the poetry of materiality and the materiality of poetry, Bök draws out the dual nature of atoms as information and matter by making his poem into an art *object*, a material LEGO book that draws upon the distinctive int(e)ractability of the LEGO System. Collapsing any opposition between conceptuality and materiality, this artistic composition actively and intimately engages the intermingling of idea and thing. Here, the poetry lies in how the material artwork functions as a medium that structures the creative repurposing of its constituent elements, as Lévi-Strauss (1966) notes of bricolage:

> Further, the 'bricoleur' also, and indeed principally, derives his poetry from the fact that he does not confine himself to accomplishment and execution: he 'speaks' not only *with* things, as we have already seen, but also through the medium of things: giving an account of his personality and life by the choices he makes between the limited possibilities. The 'bricoleur' may not ever complete his purpose but he always puts something of himself into it. (p. 21)

Speaking *with* things, poetic bricolage transcends language to dialogically engage with the world of material objects. Yet, also speaking *through* things, Bök's poetic bricolage deploys the material grammar of a systematic object language for artistic expression. In so doing, this poetic bricolage performatively explores the mediateness of bricolage itself, a performative space for creatively reassembling the already-significant interplay between materiality and signification.

Unlike how the core play cultivated by LEGO advertises its combinatorics as simply expanding the possibility space of LEGO play, Bök's poetic decomposition probes the precarity of combinatorics, interrogating how human understanding often takes the form of LEGO-like poetic composition, combining and recombining symbols into meaningful assemblages that construct rather than merely reflect reality. Furthermore, with their fragmented, permuted meanings, Bök's two artworks demonstrate that composition—poetic, material, linguistic—depends on decomposition.

This meditation on bricolage, therefore, offers a very different picture of the mixed messages of LEGO than that typically found in most corporately-authored LEGO playscripts. Rather than synthesizing otherwise contradictory messages into a hybrid media form, Bök's poetic play defamiliarizes the very act of making sense—how seemingly nonsensical arrays of material elements become meaningful in certain configurations and how seemingly sensible elements become meaningless in other configurations. This defamiliarizing, deconstructive, and decompositional play calls bricolage into question as a meaning-making process. At the same time, it also paints a compelling picture of a world in which meaning-making generally follows the compositional logic of bricolage. In the spirit of int(e)ractability, therefore, Bök's poetic decomposition demonstrates how the very lines of fragmentation drawn by making sense of the world are also the LEGO-like points of connection from which we can forge poetry and play.

Works Cited

Alsdorf, David, and Brian E. Gravel. 2018. Representational praxes. In *Making engineering playful in schools*, ed. Brian E. Gravel, Marina Bers, Chris Rogers, and Ethan Danahy. Billund: The LEGO Foundation.

Bernstein, Robin. 2009. Dances with things: Material culture and the performance of race. *Social Text* 101 (27): 67–94.

Bogost, Ian. 2007. *Persuasive games: The expressive power of videogames*. Cambridge: The MIT Press.

Bök, Christian. 2012. Ten Maps of Sardonic Wit. *MIT Unbound Symposium*. https://futurebook.mit.edu/2012/05/ten-maps-of-sardonic-wit-christian-bok/. Accessed 9 April 2020.

de Certeau, Michel. 1988. *The practice of everyday life*, trans. Steven Rendall. Berkeley: University of California Press.

Freud, Sigmund. 1989. *Beyond the pleasure principle*. New York: W. W. Norton.

Gauntlett, David, Edith Ackermann, David Whitebread, Thomas Wolbers, and Cecilia Weckstrom. 2010. *The future of play: Defining the role and value of play in the 21st century*. Billund: LEGO Learning Institute.

Genette, Gérard. 1997. *Paratexts: Thresholds of interpretation*, trans. Jane E. Lewin. Cambridge: Cambridge University Press.

Gray, Jonathan. 2010. *Show sold separately: Promos, spoilers, and other media paratexts*. New York: New York University Press.

Gregg, Peter B. 2018. Brick-olage and the LEGO/brand axis. *The Popular Culture Studies Journal* 6: 29–44.

Grobovaite, Dalia. 2017. Politics of bricolage and the double-sided message of *The LEGO Movie*. *Canadian Journal of Media Studies* 15: 57–78.

Hayles, N. Katherine. 2004. Print is flat, code is deep: The importance of media-specific analysis. *Poetics Today* 25: 67–90.

Hebdige, Dick. 1988. *Subculture: The meaning of style*. New York: Routledge.

Jenkins, Henry. 2013. *Textual poachers: Television fans and participatory culture*. Updated Twentieth Anniversary Edition. New York: Routledge.

Lauwaert, Maaike. 2009. *The place of play: Toys and digital cultures*. Amsterdam: Amsterdam University Press.

Lee, Jonathan Rey. 2014. The plastic art of LEGO: An essay into material culture. In *Design, mediation, and the posthuman*, ed. Dennis M. Weiss, Amy D. Propen, and Colby Emmerson Reid, 95–112. Lanham: Lexington Books.

Lee, Jonathan Rey. 2019. Master building and creative vision in *The LEGO Movie*. In *Cultural studies of LEGO*, ed. Rebecca C. Hains and Sharon R. Mazzarella, 149–173. Cham: Palgrave Macmillan.

A LEGO Brickumentary [Film]. 2014. Directed by Kief Davidson and Daniel Junge, Radius.

The LEGO Movie [Film]. 2014. Directed by Phil Lord and Christopher Miller, Warner Bros.

Lévi-Strauss, Claude. 1966. *The savage mind*. Letchworth: Weidenfeld & Nicolson Ltd.

McLuhan, Marshall. 2005. *Understanding media*, 2nd ed. Routledge.

Turkle, Sherry. 1997. *Life on the screen: Identity in the age of the internet*. New York: Touchstone.

Zinguer, Tamar. 2015. *Architecture in play: Intimations of modernism in architectural toys*. Charlottesville: University of Virginia Press.

Zosh, Jennifer M., Kathy Hirsh-Pasek, Emily J. Hopkins, Hanne Jensen, Claire Liu, S. Dave Neale, Lynneth Solis, and David Whitebread. 2018. Accessing the inaccessible: Redefining play as a spectrum. *Frontiers in Psychology* 9: 1–12.

CHAPTER 2

Housing Play in LEGO Construction Toys

"Playing house" may seem to more aptly describe LEGO play than the imaginative play to which this phrase commonly refers. After all, LEGO is a construction toy designed for building miniature houses, whereas "playing house" is typically more about family dynamics than architecture. Upon closer examination, however, these forms of play mirror each other. Relying on the close metonymic connection between house and family, "playing house" dramatizes family dynamics in ways that assume and imply domestic spaces. Reciprocally, LEGO's *housing play*[1] invokes domestic ideologies in the design of its architectural spaces. In these intertwined forms of play, the spatial is social and the social is spatial. Thus, while no theory of LEGO is complete without accounting for *construction play*—the hands on bricolage of physically reassembling already-significant material elements—it is also important to recognize that this is an ideological as well as a material practice. Indeed, the ideological construction of construction itself unites LEGO's many play themes. By tracing this deeper ideology of construction across LEGO's housing-, town-, and city-themed sets, this chapter aims to deconstruct a more widely domesticating ideology that has come to define the LEGO construction toy.

[1] This phrase references both how LEGO plays at housing and how LEGO houses its ideology of play in its construction toys.

J. R. Lee, *Deconstructing LEGO*,
https://doi.org/10.1007/978-3-030-53665-7_2

Never a purely abstract medium (see Preface), "LEGO always had its roots in architecture" (2015, p. v), as Tom Alphin points out in his brick-based guide to architectural styles. "After all," Alphin continues, "the little plastic pieces are called bricks" (p. v). From its rectangular bricks to its sloped 'roof tiles,' LEGO's most foundational elements have always been architectural (Fig. 2.1). To build anything with this system is to engage in an architectural logic manifested in a material process resembling bricklaying[2]—even a brick-built mountain or giraffe structurally implies construction.

Construction is embodied both in material LEGO elements and in the organizational principles behind the interconnection of said elements—after all, the latter are materially embodied in the former. As Beijing architectural firm MAD notes in their contribution to the LEGO Architecture Studio (Set #21050) booklet, "When you start to experiment with LEGO®, you can start to think of the LEGO bricks as an architectural language" (2013, p. 176). Helping to build fluency in this language, the paratextual scripts that accompany these elements—especially product images and instruction manuals[3]—teach players not only how to build a specific model but also how LEGO elements are designed to go together.

That is, implicit in the fixed procedure of building according to the instructions is a *procedural rhetoric* that encourages even the most mechanical builders to become fluent in the underlying principles of LEGO design, including both the grammar of LEGO connections and the symbolic vocabulary of LEGO elements. Players are encouraged to follow a single logic of construction whether they mechanically follow the exact instructions or creatively depart from these instructions (or some combination thereof[4]). And, as all other forms of LEGO play either involve or presuppose this foundation of construction play, the ideological

[2] Heightening this resemblance, Jan Vormann has performed an unusual style of interactive LEGO art by using LEGO to patch eroding structures around the world. In many cases, he uses LEGO bricks to fill in gaps in crumbling brickwork.

[3] LEGO also has other paratextual avenues for teaching construction principles, including building guides in the form of online videos, magazine features, and picture books and a product line of Master Builder's Academy sets (see Lee 2019) that teach building principles within the context of building a particular set.

[4] While it is certainly common to differentiate these modes of building, they are more interconnected than they may appear. As Seth Giddings (2014) finds in his ethnographic work on LEGO players, "Others, however, have found following instructions compelling and creative in its own way, the ostensible constraints on imagination affording instead an

Fig. 2.1 **a** The LEGO Company's first iteration of the brick, the 1949 Automatic Binding Bricks (Set #700-12), demonstrate 'bricklaying' as the foundational mode of LEGO construction and includes door and window elements to enhance the architectural focus. The images on the front of the box show a higher proportion of red and white than the elements shown in the box, suggesting an architectural aesthetic based on brick and plaster. This aesthetic is seen again in **b** the 1968 Basic Set (Set #066-1), which still privileges architectural construction even while expanding its range of depicted constructions to include figures and vehicles. In both these sets, note the defining features of "classic" LEGO: architectural construction primarily using rectangular *bricks* and *slopes* (the red roof tiles) accompanied by a few specialized elements like doors, windows, and wheels

construction of construction itself comes to define the LEGO medium and brand.

While the architectural foundations of LEGO are easy to see, it may be less immediately clear how LEGO continues to domesticate its architectural spaces, especially those spaces which are not thematically domestic. This chapter argues that the domestic and domesticating ideologies that script LEGO construction *themes* simultaneously inflect LEGO construction *play*. These thematic presentations generalize themselves by establishing an ethos of connectivity that applies not only to particular LEGO toys but also to the stance children are encouraged to take with respect to LEGO construction. LEGO housing play, that is, promotes a domesticating, architectural way of thinking as part of the core LEGO brand identity, thereby inflecting the medium itself as a domestic, architectural toy.

Tracing LEGO construction play from the more recent LEGO Creator (2001–present) houses back to the suburban ethos of the Town Plan (1958–1966) and forward again to the fragmented urban sites of *LEGO City* (2005–present), this chapter deconstructs the themes that most directly construct the LEGO ideology of construction play. The following case studies explore how LEGO plays off the nostalgic allure of the miniature tradition while promoting this systematic, technologized medium as a regulatory, domesticating force. Although sourced in a nostalgic postwar retreat to suburban and domestic spaces, LEGO surprisingly does not often emphasize the inhabitation of such spaces. Instead, its contribution is more ideological, playing out primarily in how LEGO scripts its architectural construction play as domesticating miniature spaces.

LEGO Houses as Domestic Spaces

To play house—in person or LEGO—is to be immersed in what philosopher Gaston Bachelard (1958/1994) calls *the poetics of space*. Drawing on the spatial metaphors that pervade ordinary language, philosophy, and poetry, Bachelard contends that the imagination operates spatially, writing that "On whatever theoretical horizon we examine it, the house image would appear to have become the topography of our intimate being" (p. xxxvi). Thus, he boldly spatializes the self, writing that "Our soul

intellectual pleasure in the process of construction, inseparable from other playful poeitic (sp.), and imaginative activities" (p. 249).

is an abode. And by remembering "houses" and "rooms," we learn to "abide" within ourselves" (p. xxxvii). While profound, this sentiment is colored by Bachelard's focus on the contemplative arts of philosophy and poetry. Considering that the house shelters *play* as much as "the house shelters daydreaming" (p. 6), one might reconstrue this relatively static sentiment as something more dynamic: *our selves are built environments, and by playing at houses and rooms, we learn to reconstruct ourselves*.

At the same time, the transformative potential of such imaginative housing play is shaped by the architecture of a construction toy's embedded playscripts. For LEGO to become a tool for speculatively reimagining the social construction of domestic space, its implicit domesticating ideology must first be deconstructed. This entails resisting how the core LEGO architecture is designed to domesticate ostensibly abstract construction play by playing out a fairly conventional social construction of domestic space. Following a burgeoning postwar ideology of domestic spaces as sheltering childhood and children's play, LEGO was initially designed according to the domesticating imperatives of "healthy, quiet play" and "indoor play"[5] (Baichtal and Meno 2011, p. 14). And while actual LEGO play may be more chaotic, destructive, and unruly than one might expect, the medium and its paratextual scripts ideologically construct LEGO play as ordered, constructive, and systematic.

Ironically, despite domesticating space being central to LEGO construction play, actual domestic spaces—i.e. houses and their interiors—make up a surprisingly small minority of its contemporary sets. Even more surprisingly, they are rare even within LEGO Architecture (which mostly focuses on monumental or urban architecture)[6] and LEGO City (which mostly focuses on urban construction and protection, as discussed later in this chapter). Instead, actual LEGO houses are currently most often

[5] This imperative was strong enough that, for a time, LEGO instruction manuals opened with a pictogram suggesting that LEGO could not be used on grass.

[6] Despite a similar reliance on architecture, LEGO Architecture is a different kind of theme from the housing play explored in this chapter. Aimed at adults, LEGO Architecture sets primarily offers collectable display models of famous architectural sites from around the world. The sets themselves feature more muted tones than is typical for LEGO and the product packaging is sleek, black, and clearly targeted at adults. Whereas many LEGO structures are roughly ***minifigure scale*** (a size that minifigures can occupy), many LEGO Architecture sets are much smaller scale. Combined with a common motif of plans and blueprints, LEGO Architecture treats architectural sites as instances of aesthetic design rather than occupiable spaces—they are more models than miniatures.

Fig. 2.2 This 1973 postcard from the short-lived LEGOLAND theme park in Sierksdorf, Germany depicts a family viewing a miniature replica of an English town rendered in LEGO. Playing with inside and outside, this image places the viewer inside the walled-off miniature city, looking outward as the tourist family looks in. Rather than advertising the LEGOLAND tourist experience, this image instead advertises the experience of the miniature sublime. Furthermore, the postcard connects the miniature experience to that of the *souvenir*, another of the cultural practices Stewart (1984) compares in *On Longing*. The LEGO Architecture line emerged similarly: Adam Reed Tucker's first designed set was initially marketed as a localized miniature souvenir sold in nine Chicago gift shops (Robertson 2013, p. 208)

found within the LEGO dollhouses of the 'for girls' Friends line (see Chapter 3) and the idyllic cabins and suburban homes of the Creator line.[7]

When LEGO does represent literal domestic spaces, its houses offer a miniature perspective on domestic architecture (Fig. 2.2). Despite or even

[7] This is especially significant because the 'theme' of the Creator series is ostensibly not a theme but a process: creation or creativity. These sets advertise themselves as throwbacks to the ideal of more abstract and creative block play that is commonly associated with traditional LEGO and yet generally have well-defined play themes.

because of its size, the *miniature* has loomed large in Western material culture and theory. No matter how inviting LEGO worlds may seem, their scale precludes the player by creating spaces too small to be inhabited by humans.[8] As Steven Millhauser reflects, "is it possible that the deepest fascination of the miniature lies here, in the unfulfilled yearning to be part of that world? For we are in disharmony with the world, we do not fit in anywhere" (1983, p. 135). Similarly, the deepest fascination of the LEGO house is that a medium that is incredibly open to construction, manipulation, and play is nonetheless perpetually closed to inhabitation. Indeed, the fact that miniature domestic spaces offer no possibility of domicile only intensifies their ideological promise—the fascination of an idealized domesticity.

As a world of our own creation that has no room for us, LEGO also plays at "architecture from the outside," an exercise in perspective presented by philosopher Elizabeth Grosz:

> The outside is the place one can never occupy fully or completely, for it is always other, different, at a distance from where one is. One cannot be outside everything, always outside: to be outside something is always to be inside something else. To be outside (something) is to afford oneself the possibility of a perspective, to look upon this inside, which is made difficult, if not impossible, from the inside. (2001, p. xiv)[9]

Applying this to the miniature perspective, the play of inside and outside becomes a wavering between two unstably linked spaces. The inside cannot be occupied because the fascination of the miniature depends upon the sense of idealized inaccessibility that emerges when the gaze reaches into spaces too small to admit the body. Yet, the outside also cannot be occupied because, as Grosz notes, the outside is always inside something else. Thus, LEGO situates the player as at once outside the

[8] LEGO playsets remain squarely in a miniature scale although it is certainly possible to build bigger. Most notably, British television personality James May commissioned an entire habitable house of LEGO, proving that such a feat is both possible and exceptionally impractical.

[9] Bachelard also complicates the distinction between outside and inside, arguing that it reifies a problematic metaphysical binary as "Outside and inside form a dialectic of division, the obvious geometry of which blinds us as soon as we bring it into play in metaphorical domains. It has the sharpness of the dialectics of *yes* and *no*, which decides everything" (1958/1994, pp. 211–212).

domestic spaces of the LEGO world and inside the domestic space of the home with its sheltered, quiet, indoor play.

The "possibility of a perspective" that LEGO offers can, therefore, be considered a kind of *miniature sublime*. Whereas the Kantian sublime depends upon the contradictory sensation of perceiving something (say, a mountain) as beyond human scale yet fitting into human conceptuality or perception, this miniature sublime depends upon the contradictory sensation of perceiving something as too small for human scale yet evoking a utopic ideal of domesticity that seems to transcend human agency. Consequently, this tension between inside and outside only reinforces the idealization of the inside as exemplifying the protectedness that defines domestic spaces. The player experiencing the miniature sublime may be outside miniature LEGO houses, but is always looking inwards toward its domestic ideal.

This sense of the miniature sublime is only intensified by the performative process of LEGO construction play. As LEGO architecture is literally constructed "from the outside," builders wall themselves off from their miniature creations with each brick they place. Thus, the instruction manual will often reveal interiors which are easily accessible during the process of their construction and much less accessible after the fact. Indeed, this is precisely where the tiny instruction manual page depicted in Fig. 2.3 is opened—with the structure largely built except for an absent roof, allowing the downward angle of the implied gaze to penetrate straight into the cramped interior of the treehouse. Thus, the miniature sublime of the finished set is intensified by the nostalgic resonances of a physical artifact that functions at once as a miniature house and as a tangible reminder of its own more accessible construction.

This interplay of inside and outside is further illustrated by the 3-in-1 LEGO Creator Treehouse (Fig. 2.3) itself, a set of three wilderness-themed domestic spaces that all funnel the gaze from the outside toward the inside. This ambiguity is characteristic of an architectural medium that often suggests a separation between these spaces—although the nature of the plastic LEGO medium is such that there is no material distinction between architecture and nature. Each of these three sets plays with the permeability of build environments by featuring—instead of doors—openings that create continuity between interior, domestic spaces and exterior, natural spaces.

The treehouse and barn feature a common design that opens the fourth wall to the player (much the way sitcom sets or theater spaces

Fig. 2.3 **a** The box of the 2013 LEGO Creator Treehouse (Set #31010), which also includes instructions for rebuilding the same set of elements into a "lakeside hut" or "farmyard barn" (depicted in the panel on the bottom of the image). Note that the red slope bricks that were introduced as 'roof tiles' in the early days of LEGO are still used to construct roofs in all three sets. Embedded in the "3 in 1" logo is a tiny graphic (indicated by arrow) of the instruction manual for this exact set. **b** The left page from the instruction manual depicted in the graphic. Note that most of the wall has smooth LEGO plates on top, leaving only a few studs for the roof section to snap onto. This is a common technique for weakening LEGO bonds so a roof can be easily removed, thereby reconfiguring the play of accessibility and inaccessibility that anchors the miniature sublime

open up a fourth wall to the viewer, as discussed in Chapter 3). In the treehouse, the walls of the treehouse, the trunk of the tree, and a low railing together bound a playable area that is at once interior and exterior. This plays out very differently in the lakeside hut, which is a fully enclosed space—one which, however, relies on two liminal spaces: the landing that extends directly from one opening in the building to become a dock and the staircase that extends directly from the other opening. Although these are ostensibly sites of transition, they form sizable playable spaces in themselves. After all, surfaces and exteriors are where LEGO studs—and thereby, interconnective potential—lie. Thus, the use of LEGO baseplates as building environments that invariably extend beyond the bounds of the LEGO structures implies a continuous relationship between the exterior LEGO world and the interiority of the indoor spaces that shelter LEGO play.

Indeed, LEGO architecture characteristically emphasizes accessible transitional spaces over inaccessible interior spaces, playing with permeable boundaries between interior and exterior. These three sets all depict natural spaces in various stages of being domesticated, as the natural elements all use their inherently architectural form to organize the space according to the inhabitable logic suggested by the partial domestic structures. Natural or architectural, all the elements of these sets are structured according to the unifying domestic logic of the built environment. Consequently, the tension between presence and absence, inclusion and exclusion, define the construction of domestic spaces in these LEGO sets and, by extension, the domestication of space throughout the LEGO medium.

Thus, domestic spaces in LEGO never feature the kinds of copious, accessible interior space often found in traditional dollhouses. In contrast, LEGO's domestic spaces mediate the inside through the outside, visually and tangibly inviting players to play in the transitional spaces that surround and frame domestic interiors. Similarly, most LEGO housing play privileges exteriors over interiors. As the following explorations of the LEGO Town Plan and LEGO City will demonstrate, this ideal of domesticity orients LEGO's ideological presentation of town and city environments, despite and even because of their surprising lack of accessible interior domestic spaces. The exteriority of these town and city spaces, that is, reinterprets the miniature sublime. Here, the intimate, inaccessible interiority underlying the fascination of the miniature is broadened to encompass the wider scope of regulatory, public, construction play.

The Town Plan

Although LEGO can be considered a medium for playing house, its earliest incarnations were more explicitly oriented toward playing *town*, specifically the *suburban* town. The earliest LEGO sets were released as part of a Town Plan (1958–1966) whose miniature suburban sprawl played out many of the constitutive paradoxes of suburban culture in ways that play with the profound paradox of the miniature sublime. Before ever establishing interior domestic spaces, the Town Plan played out suburban anxieties surrounding the domestication of the suburban sprawl itself. That is, it played out the uneasy collision between a nostalgic ideal of sheltering and nurturing and progressive ideal of technological development that motivated a suburban obsession with traffic regulation.

While LEGO players can be reasonably assumed to include urban and rural citizens, the domesticating impulse behind LEGO construction play is closely aligned with suburban ideals. Historically, LEGO was first introduced at a time when the flight to the suburbs was transforming the Western cultural landscape—not only by providing viable places to live, but also by offering dreams of a kind of utopic space discursively connected to the ideals of domesticity, childhood, and play. As Maaike Lauwaert notes, "the suburb became itself an epitome of the postwar consumerist culture" as a space that "was in essence child-centered" (2009, p. 31). The suburban ideal, that is, primarily located its sense of domestic bliss in sheltering the futurity represented by childhood, including the increasing cultural valuation of creativity and play.

Recovering from the horrors of world wars in part by promoting nuclear families, postwar culture was in many ways an ideal market for a modern educational toy like LEGO. The scripted yet creative play inherent in the LEGO System fit with the fusion of conformity and creativity that Amy Ogata describes as defining this period. Ogata notes that even within stereotypically cookie-cutter suburban spaces, a maker culture arose that "made do-it-yourself home repair, construction of furniture, and creating children's amusements the expectation of middle-class homeowners" (2013, p. 99). As this maker culture aimed to develop and maintain child-centered domestic spaces, increased cultural attention also began to be paid to the objects that populated such playspaces. For instance, a still-ongoing tension arose as "ambitious parents, wanting to maintain a wholesome distance from 'lowbrow' mass consumption, favored toys that evoked traditional craftsmanship while also hoping to satisfy their children's desire for entertainment" (p. 9). In many ways, LEGO was perfectly poised to offer its Town Plan as an entertaining-yet-educational toy that both mimicked and fit into the evolving suburban maker culture.

In other words, the Town Plan spoke to a suburban ethos whose "two foci" were "the nostalgic and the technological" (Stewart 1984, p. 1). At once looking backward to a nostalgic vision of domestic tranquility and looking forward to a technologized future where automation frees the domestic, utopic suburban spaces sought to leverage the latter to restore the former. In this ideological construction, the modernity of suburban spaces lay in using advanced design principles to organize and regulate domestic spaces and domestic labor, thereby sheltering and freeing domestic life.

LEGO design appealed to several elements of this nostalgic-yet-technological ethos. As a modern medium, LEGO's paradoxical fusion of traditional wooden blocks and modern plasticity[10] provided a nostalgic-yet-technological material foundation for construction play. Then, as a modular construction system, LEGO provided a rational, systematic approach to architecture that organized the playful design and construction of domestic spaces. And more broadly, the Town Plan itself functioned as a unifying system for regulating the suburban sprawl as a material toy system that exemplified the regulatory ideals that already governed suburban design. Whereas the regulatory ideal applied to interior, domestic spaces centers on the authority of parents as simultaneously sheltering and disciplining their children, the regulatory ideal of the suburban sprawl centers on a fundamental design logic—the rational organization of domestic spaces into planned communities.

[10] Despite their technological aura, plastics (like automation) were seen as supporting domesticity. As Lauwaert notes, "The relatively cheap, easy to clean, durable, colorful and standardized plastic bricks epitomized the postwar vogue of indoor plastic products" (2009, p. 58).

a

Fig. 2.4 **a** Danish Catalog, 1958. This grid of product images depicts the available numbered sets as elements in a collection. The two larger images depicting these products spread out on the available playmat then show how this collection can be unified to form a Town. Five of the sets in the upper left quadrant are vehicular-focused. Note the playmat featured prominently in the lower-left quadrant behind the cartoon of a child builder whose design prefigures the red cap and overalls of a common early minifigure

Anchoring this collectivity is a now largely defunct[11] element: the *playmat* (Fig. 2.4). Consisting of a printed cardboard or wooden slab with printed roads and plots, this playmat exists outside the LEGO System

[11]Versions of the playmat have appeared intermittently across LEGO history, although not generally as the organizing principle behind an entire product line as in the Town Plan. Most notably, LEGO baseplates offered large studded building surfaces that combined elements of LEGO bricks and playmats, drawing the playmat into the medium. In the late 1970s a series of 32x32 smooth-bottomed baseplates brought back a road system akin to the Town Plan. In the 2000s, third-party producers created non-LEGO playmats as ways to organize LEGO Robotics competitions. And more traditional playmats officially returned in 2012 with revamped cardboard playmats featuring the layout of Heartlake City from LEGO Friends. Playmats (including City-themed ones) continue to be sold but are not generally advertised or featured within LEGO product lines.

b

Fig. 2.4 **b** An image showing the playmat in greater detail from the 1961 Town Plan—Continental Europe set (Set #810-2). The localization of this product emphasizes the subtle details of the regulatory play patterns with all vehicles driving on the proper side of the road and obeying the road signs and traffic cops. A localized UK and Australian version of the LEGO Town Plan Board (Set #200-5) from 1962 shows an identically-framed product image (two different children in the exact same angle and pose, both with the boy leading the regulatory play while the girl spectates) with the vehicles on the other side of the road, emphasizing the importance of paratextual playscripts in relating regulatory norms (after all, the roads and playmats have no inherent directionality)

(it has no studs) yet exemplifies systematicity by presenting a topography that connects sets into a single town. A material manifestation of town planning, the playset 'grounds' planned communities that bear a reasonable resemblance to suburban environments in which spatial proximity unifies residents into a community.[12] This reveals a suburban ethos characterized by a chain of nested part-whole relationships: the suburban community shelters its nuclear families just as rationally organized towns shelter their domestic spaces. Thus, the extreme insularity of the single-family home is maintained within a larger network of communal relations that define the suburb as a space of safety analogous to the family homes they shelter. In safety if not intimacy, suburban communities are idealized as extended families that share responsibility for maintaining the sanctity of the suburban space.

These emergent suburban communities, however, are anything but nuclear. Cities have centers—nuclei that serve as hubs for radial and radiating circles of activity. Suburbs, by contrast, have centerless grids that regiment their largely homogeneous rows of houses. Roads unify and connect the suburban spaces they organize, defining neighborhoods and blocks and rendering communities accessible. It is appropriate, therefore, that roads constitute by far the largest and liveliest expanses on the Town Plan playmat. Although one might expect the Town Plan to star the LEGO construction toy, the exterior-focused structures in the Town Plan lack accessible interiors or notable play features. As these manifestations of suburban architecture rest passively on the playmat in all their miniature splendor, play turns instead to the detailed suburban infrastructure.

Inhabiting the network of roads and crosswalks are the mobile figures of the Town Plan: a series of vehicles, figures, and moveable road signs that bring a sense of dynamic motion into LEGO's miniature worlds. Toy cars and trucks dominate the scene, providing all the animating force depicted in Fig. 2.4. The primacy of the vehicle in this suburban drama is only reinforced by the occupations of the occasional human figures such as traffic cops (the first figures in the LEGO world), cyclists, and pedestrians. These inhabitants—the vehicles and figures alike—represent generic flows and potentiality for movement, nested as they are within these traffic networks. That is, these relatively faceless agents—the vehicles synecdochically standing in for the citizens they mask—are treated less as individual subjectivities and more as participants in the communal dance of suburban movement.

This enacts a striking reversal in which the most dynamic aspects of the Town Plan fall outside the LEGO medium. The vehicles and figures are

[12] Something similar occurs at the national level when geopolitical borders begin to define national identity, as Benedict Anderson (2000) argues in *Imagined Communities*.

pre-molded, non-articulable toys with no literal or figurative connection to the LEGO medium (quite unlike the minifigures that debuted about two decades later). The roads on the playmat are not only not LEGO, they are mere surfaces—empty gaps between built LEGO structures that gain significance primarily through printed textual and contextual cues. The elements that unify and animate the Town Plan, that is, paradoxically also render the LEGO medium and its architectural playsets as static and immobile backdrops to suburban play.

In this way, the Town Plan mirrors an internal tension characteristic of suburban spaces, namely the simultaneously complementary and contradictory relationship between roads as sites of dynamic (and dangerous) motion and houses as islands of domestic tranquility. Lines of both disconnection and connection, roads are the primary way suburbs domesticate their domestic spaces, serving as boundaries as well as pathways. Roads make the suburbs possible since suburbs privilege domestic space to such an extent that they are uninhabitable without reliable supply lines. And yet, roads thoroughly contradict the suburban idyll: technological rather than nostalgic, functional rather than aesthetic, synthetic rather than natural, fast rather than slow, and—most pressingly—dangerous rather than sheltering.[13]

A concern with traffic safety is, therefore, a core element of the suburban ethos and a primary motivating force behind the design of the LEGO Town Plan. Emerging from a postwar Danish culture in which "children tend to equate traffic with automobility: Cars are seen as harmful objects that the children should avoid at any cost" (Kullman 2009, p. 208), the Town Plan was "produced in collaboration with the Danish Road Safety Council" such that "the sets helped teach traffic safety to children in an era when automobile ownership was steeply on the rise" (Lipkowitz 2012, p. 19). And even while more recent play themes have become significantly more violent, LEGO has never fully abandoned its concern with road safety. For example, many contemporary LEGO media texts include recurring seatbelt jokes (most notably the *LEGO Batman* movie), the characters from *The LEGO Movie 2* starred in an airplane

[13] More specifically, the paradox of suburban car culture is that the very transportation system that connects the suburban parent to the urban workplace undermines the point of the suburbs (to insulate the child from urban traffic). This technological necessity thereby threatens the nostalgic ideal it makes possible.

safety video, and LEGOLAND continues to run a longstanding miniature traffic school[14] in which children can earn LEGO driver's licenses for following safety principles.

Even as the Town Plan privileges vehicles as the primary inhabitants of its suburban spaces, it embraces a socializing message of vehicle safety, scripting play toward the regulation of traffic. Rather than avoid the dangers of car culture, whose technologically-enhanced mobility threatens the safety of the unmediated body, the Town Plan subsumes this technological danger under the nostalgic ideal of fully regulated suburban spaces. The Town Playmat is structured by a clearly defined grid of roads and crosswalks that both offers regulatory playscripts and spatially separates the vehicular playspace from the domestic playspace. And the green lawn spaces that narrowly ring many of the building plots arguably function more like buffers than playspaces. The playmat, that is, is predicated on drawing boundaries between static domestic spaces and a motive transportation network that is bounded and regulated by the circumscribing presence of the domestic spaces.

The Town Plan further domesticates its vehicular play by providing its vehicles their own 'domestic' spaces where they can be housed and cared for, including garages, a filling station, and a service station. Indeed, the sets featured in Fig. 2.4 have a much wider variety of vehicular spaces than human-centered ones. They also feature much more accessible designs with external bays to house cars and even a sophisticated garage door mechanism (Set #236, Fig. 2.4). Thus, as they are re-placed in miniature, the vehicles of Town Plan are transformed from dangerous technological incursions to be avoided to docile agents to be regulated and cared for.

This regulatory play is one way the suburban ethos of the Town Plan subsumes the technological under the nostalgic. Rather than invite players inside miniature toy architectures, the Town Plan domesticates its exterior, public spaces. Interestingly, the Town Plan imagines regulated outdoor play from within indoor playspaces that in fact insulate the child from the outdoors. Similarly, the child accesses miniaturized public spaces through private play. Consequently, the miniature toys of the Town Plan play at domesticating exterior spaces, forging a suburban ideal in which entire planned, cultivated, and regulated suburban communities function

[14] Although Kullman (2009) does not discuss LEGOLAND in particular, Kullman does perform a spatial analysis of a somewhat similar institution called Children's Traffic City.

like fully protected domestic spaces. In the technologized nostalgic of this suburban ideal, regulatory forces create spaces of mobility without danger.

Yet, these utopic suburban spaces are only possible if people live up to the regulatory ideal implicit in their design. The suburban ideal hinges on the shared ideological framework of its residents. Similarly, the utopic design of the Town Plan hinges on players playing out the regulatory ideal offered by the Town Plan's playscripts, the willingness of children to follow its implied safety narratives. Thus, the Town Plan invites complicity in its specific suburban ethos just as the domestic LEGO medium invites complicity in its broader philosophy of play. Far from being non-ideological, the 'rationality' of its regulatory design is precisely the foundation of the Town Plan's ideology.

The Town Plan clearly invites players to participate in this regulatory thinking. Yet, despite this product line's suggestive name, players are not situated as *town planners* themselves—after all, the network of roads that constitutes the Town Plan is fixed to a static playmat and unavailable for redesign. Instead, players are invited to *construct* the town by piecing together its structures and *animate* the town by dramatizing regulated town life. Rather than constructing the regulatory system itself, players are invited to implement and play out the regulated social life of the suburbs.

While children will certainly not always straightforwardly follow these scripts, the mere presence of these scripts circumscribes children's play in a meaningful way. Whatever children actually do with them, these scripts send a strong signal to parents that the Town Plan promotes a suburban ethos. The linked messages of the Town Plan's suburban design and LEGO's promise of quiet, indoor play position the child as a *docile body* (Foucault 1975/1995) whose 'goodness' is characterized by willing obedience to the social norms of the state and family.[15] This positioning draws an analogy between the citizen's adherence to the national law and the child's adherence to an idealized form of "healthy, quiet play" (Baichtal and Meno 2011, p. 14). The emergent play culture reflects what Michel Foucault (1975/1995) calls a *society of discipline*, a system in which regulation is primarily self-enforced by people who are socialized to uphold the rules of the system.

[15] See Brian Sutton-Smith's account of how toy-giving rituals, especially around the commercialized Christmas traditions, act as a kind of social contract that use toys as both rewards and *methods* for docility (1986, pp. 20–21).

And even when children deliberately flaunt these regulatory rules by violating the scripted traffic patterns, in so doing they may actually be demonstrating developing fluency with the system. These toys promote safety not necessarily by requiring that all dramatized play in the toy world follow safety protocols, but by building fluency with the protocols. Rather than being genuinely transgressive, playing out violations may be a way of learning the regulatory system as these toys promote fluency with reading the road markings, traffic lights, and signs that weave imperative messages into the material design of the road system. While there are certainly good reasons to promote traffic safety in children, it is important to recognize that the practical skills taught here are just a few more obvious manifestations of a complex suburban ethos that has significant socializing tendencies.

As with any socializing toy, the Town Plan teaches children to decode social signifiers and, in so doing, take up the cultural ideologies that underlie them. This kind of socialization contributes an imperative toward development itself reflective of the suburban imperative to shelter childhood. Here, the child is not only a figure of innocence (being-child) to be protected but also—for good and ill—a figure of futurity[16] (becoming-adult) to be educated. The implied future of the developing child is, moreover, temporally analogous to the spatially implied cities suggested by the Town Plan's roads and vehicles. Presenting neither childhood nor the suburban spaces that shelter childhood as fully self-sufficient, the Town Plan disrupts its suburban idyll by privileging a traffic network that perpetually points to its absent yet implied shadow self—the fast-paced, grown-up world of the city.

LEGO City

Over several decades, the suburban design of the Town Plan gave way to a series of town- and city-themed lines that have become increasingly urban. The current incarnation, LEGO City (2005–present), encompasses a range of sub-themes that mimic elements of urban environments, including construction, mining, police, fire, search and rescue, and space

[16] See Lee Edelman's (2004) theory of futurity in *No Future* or Elizabeth Freeman's (2010) theory of chrononormativity in *Time Binds*.

exploration.[17] This represents a significant shift from the suburban Town Plan, replacing a paradigm of the town as spatially organizing and regulating a domestic ecosystem with a paradigm of the city as site and incitement to urban drama, a paradigm which in many ways resembles Marc Augé's description of the *non-places* of supermodernity—spaces defined by the movement of the traveler through "the urgency of the present moment" (1992/2008, p. 104). In opposition to the utopic spaces of the spatially-regulated Town Plan, the non-places of LEGO City operate as settings or stages for implied encounters and are, therefore, as much textual (and contextual) as architectural.

As how LEGO playsets stage *dramatic play* is the topic of the following chapter, this section traces how LEGO City both dis-places and re-places the suburban ethos of the Town Plan by deconstructing its spatial organization while maintaining much of its ideological orientation in its three most dominant sub-themes of *construction*, *protection*, and *transportation*. I suggest that by presenting urban landscapes as spaces of movement and action, LEGO injects an air of excitement into its "healthy, quiet play," thereby modernizing a nostalgic medium without overturning its nostalgic elements.

The primary point of departure for the more modern LEGO City lines is the absence of the playmat as a unifying spatial principle. Whereas the spatial continuity of the Town Plan playmat organized traffic patterns to present suburban spaces as domestic ecosystems, LEGO City depends upon the agency of minifigures and vehicles to draw *implied* connections across spatially disconnected locations. Ironically, there are no full-fledged urban environments in LEGO City—no city blocks lined with shops, office buildings, and high-rise apartment buildings. Instead, in a key departure from Augé's theory, LEGO City represents spaces of supermodernity from within a self-reflexively postmodern miniature toy frame that persistently emphasizes its perpetual deconstruction and reconstruction. Thus, one of the central events of LEGO City play is the reconfiguration of its own form.

[17] This space exploration adopts a realistic NASA-inspired theme rather than a science-fiction aesthetic. This positions this sub-theme as more educational in the vein of other partnerships between LEGO and NASA, including the user-submitted LEGO Ideas set NASA Mars Science Laboratory Curiosity Rover (Set #21104) and educational uses of LEGO on the International Space Station.

By abandoning any system that would spatially organize the environment, LEGO also abandons any sense that urban planning could function as a regulatory structure like the suburban design of the Town Plan. Instead, the fragments of the LEGO City represent spaces or even times of urban drama that are connected by narrative threads rather than spatial relations. Compare, for instance, the Town Plan to the City Square set (Fig. 2.5), which eliminates the organizing principle of the playmat to instead offer an abundance of disconnected playable encounters. This flips a paradigm that privileges network over node to a paradigm that privileges node over network. Or, in the language of Ian Bogost (2006), LEGO City eschews *system operations* in favor of *unit operations*. Functionally

Fig. 2.5 Promotional image for City Square (Set #60097). Note the haphazard placement of the disconnected elements of this set, disrupting any sense of an urban grid. All modes of transportation crisscross in an undefined field in and around the four structures. Small vignettes, such as the woman feeding her dog a sausage, suggest the ceaseless motion of urban life as the citizens of this scene are almost universally in-transit or mid-action (none of the fourteen minifigure characters are sitting despite there being many places to sit including all the vehicles and the two seating areas). Even the food and drink purveyors seem to be in mid-transaction despite not having any customers. All told, what is depicted here is an idea of city life—the non-places of supermodernity—rather than any comprehensible urban space

speaking, LEGO City replaces the organizing principle of the playmat (network, system) with increased interactive and representational design in its structures and vehicles (node, unit).

In addition to fragmenting the spatial organization of urban spaces, LEGO City blurs the architectural distinction between inside and outside by replacing closed miniature models with open playsets. Whereas the closed structures in the Town Plan functioned as nodes in a network of interconnected traffic routes, the structures in LEGO City expose their interiors to an amorphous expanse unregulated by any systematic urban plan. Thus, these interior and exterior spaces merge not only with each other but also with the domestic playspace. Whereas the bounded playmat and closed architecture of the Town Plan rendered players perpetually on the outskirts of the suburban space, the unbounded playspace and open architecture of LEGO City invite players to enter into the miniature world.

To accomplish this degree of openness, however, LEGO City could never become a "City Plan" analogous to the Town Plan. Without a unifying spatial principle, LEGO City placed greater emphasis on the significations built into the sets themselves. As the medium has evolved to include many more diverse and detailed elements, LEGO City reimagined urban spaces less as spaces and more as assemblages of exaggerated signifiers. As Mark J. P. Wolf argues of LEGO *Star Wars*:

> … the LEGO versions are caricatures of the films' locations, simplifying them and exaggerating their salient features. Design elements from the overall architectural style and color palette to such things as the shapes of the light fixtures, windows, doorways, and control panels, all evoke a feeling similar to the original, yet their disproportionate sizes give them greater emphasis and make the viewer more aware of the style they represent and embody. (2014, p. 27)

In this vein, the City Square LEGO store features a relatively shallow[18] façade such that the entire structure is essentially a single window display

[18] Empty space—the space that allows movement—is unilaterally diminished in LEGO's interior designs (across all product lines). Although such structures are ostensibly built in "minifigure scale," this can be somewhat misleading as interior spaces are always significantly more cramped for minifigures than their real-world analogs. This collision of scales erodes the sublimity of the miniature perspective, which depends on the visible separation between two incompatible environments.

featuring a prominent LEGO logo, a register, and LEGO toys that are not scaled down at all. Consequently, this structure functions less to replicate the experience of occupying the real space of a LEGO story and more to replicate the associative links that define the story as a transactional, transitional space of urban encounter.

Completing the move from atop to inside the LEGO world, LEGO City also replaces the non-LEGO vehicles and figures that served as the inhabitants and agents of the Town Plan with brick-built vehicles and figures. Extending construction play from LEGO structures to the entirety of the LEGO world, buildable vehicles and figures suggest that urban life, as well as urban spaces, are subject to construction. This radically transforms many of the messages of the Town Plan. Whereas the Town Plan's car-centric sets offered the *idea* of vehicle maintenance to symbolize the domestic care of its vehicular inhabitants, City Square's car-centric structures offer to play out actual maintenance as the included tools and workable automotive lift only symbolically reference the possibility of reconstruction and repair already built into the LEGO vehicles themselves.

Further extending this emphasis on reconstruction is a prevalent subtheme not shown in City Square, namely construction equipment. Far outstripping the number of domestic spaces in the line, construction equipment sets present urban 'spaces' not as spaces but as implied sites of perpetual reconstruction. A particularly telling example is a series of demolition-themed sets such as the 2015 Demolition Site (Fig. 2.6), which features three detailed yellow construction vehicles and a shabby condemned building consisting only of a few breakable walls. This rare inclusion of an urban domestic space emerges as a sign of its erasure. Too ugly to nostalgically represent a bygone domestic idyll, this ready-for-disassembly architectural frame instead presents the timeless nostalgia of this idyll as an obstacle to an ongoing drama. Like a shark that cannot stop swimming lest it die, this vision of urban life is animated by a persistent imperative to never stop reconstructing itself. The animating force of the LEGO City is, therefore, the cycle of its perpetual erasure and rebirth, deconstruction and reconstruction.

In this urban drama of perpetual reconstruction, buildable vehicles and figures return not only as material embodiments of the possibility of reconstruction but as active agents in reconstruction. Whereas the vehicles and figures in the Town Plan functioned as inhabitants of suburban spaces to be animated within the motive flows of traffic patterns, these

Fig. 2.6 LEGO City Demolition Site (Set #60076). This scene is particularly dynamic with the wrecking ball shown in mid-swing, a moment of action enhanced by computer-generated motion and impact lines. This scene has an unusual temporal structure described by McCloud (1994). Although the scene ostensibly shows one slice of arrested time, the spatial juxtaposition of the two smaller vehicles suggests a temporal sequencing in which movement from right to left shows the hauling away that follows the moment of demolition. Similarly, the dynamite is shown with a lit fuse alongside a computer-generated explosion, suggesting another before-and-after moment in a single frame. Also of note, the backdrop shows an urban skyline quite clearly in the process of construction, suggesting that this demolition of suburban domestic space is a precursor to urban reconstruction

characters function here as agents of reconstruction. For instance, many of the vehicle sets build scripts into their design by emphasizing vehicle agency with moving parts like working winches and deployable ladders on rescue vehicles and all manner of cranes, scoops, and drills on the ubiquitous construction equipment. Even immobile elements often take on an agential character, as when a police car's sirens and emblems materially signify the authority to exert regulatory force, symbolically making the vehicle an agent of the state and its laws. Similarly, the increasingly representational design of LEGO minifigures and minifigure accessories provide a set of symbols that script the deployment of the minifigures' interactive potential. That is, vehicle and minifigure designs come prescripted to determine their agential role within a world that—through its increased emphasis on the playset—consists of partially opened spaces that are designed as settings for vehicle and minifigure-oriented play.

As the urban architecture implied by this theme of reconstruction is largely absent from LEGO City, the structures that do appear take on a quite different character. The relatively rare[19] structures in LEGO City are typically institutional spaces like police stations which function as hubs of activity or bases of operations (nodes) rather than dwelling places (except in the sense in which imprisonment constitutes 'dwelling'), further situating the architecture of this city as oriented toward agential and dramatic play. These sets cement *protection* as another central sub-theme of LEGO City. As LEGO City reorients its play around the urban drama, it replaces the spatial regulatory force of the Town Plan with a more mobile, action-packed regulatory drama of protection. Police, firefighters, rescue workers, and the like counterbalance the implicit danger of the city. Like the roads which paradoxically both connected and threatened suburban spaces, this theme of ongoing protection construes implied urban spaces not as spaces of total safety but of regulated danger.

Although this play at protection against danger lacks the spatial continuity of the Town Plan, its fragmented sets nonetheless manage to imply a spatial dichotomy between inside (security) and outside (danger). Although LEGO City has no geographic outskirts, it thematically evokes this dangerous fringe by using forests, mines, and criminal hideaways as spaces that threaten the integrity of the city from the outside. Although these spaces are often implied rather than represented, their palpable presence in set design presents the city as permeable, a space of potentially limitless and sometimes dangerous encounters that form the basis for the urban event as dramatic play. Returning to the example of the Demolition Site (Fig. 2.6), one can observe a ruined wooded fence separating the dilapidated structure from the perpetually reconstructed cityscape, lovingly rendered in the electric blue often associated with

[19] A search of the online LEGO Shop (December 2015, the same year as the Demolition Site) yielded thirty-five *LEGO City* sets categorized as 'vehicles' and only two (*Swamp Police Station* and *Fire Station*) categorized as 'buildings.' Both of these buildings include multiple vehicles (e.g. a fire truck, van, and helicopter in the *Fire Station* set).

modern tech culture. Visibly outside the unreachable and thereby utopic urban city, the outskirts represent an inherent threat regulated not by the flimsy background fence (which has a sizable gap) but by the protective-yet-destructive force of the mobile, vehicular demolition team.

Like the outskirts, suburban spaces are implied but not represented in LEGO City. Implicitly connecting urban playsets to absent domestic spaces, a vehicular logic of *transportation* functions as a third major sub-theme of LEGO City. Despite lacking roads, the City Square set has no lack of transportation options, featuring a train, helicopter, tow truck, delivery truck, three cars, a scooter, a bicycle, a hot dog cart, and a handcart. This jumble of fragmented connective technologies leverages an explosion of mobility to counterbalance LEGO City's erosion of space. This abundance of transportation options implies a multiplicity of spaces—especially absent domestic spaces—far away from the depicted city center.

As the four unfixed buildings are connected primarily by this extensive transportation network, the titular city 'square' presents the city as an amalgam of experiences rather than an inhabitable space—the player is positioned here not as mayor (as in SimCity or the Town Plan) but as tourist or traveler. Thus, this set brings the suburban child to the city, emphasizing the locations that would be most significant on a day trip. For example, the coffee stand and hot dog cart are oriented toward pedestrian consumption, presenting the city as a space of *flânerie*—purposeless strolling—for adults and children alike. And, of course, the whimsically self-reflexive LEGO Store anchors this sense of urban exploration as commercialized entertainment. For the LEGO-obsessed suburban child, a trip to the LEGO Store imbues an urban space with special significance, and the image of that destination may well structure the child's hazy sense of urban geography.

These fragmented yet associated images reflect how Augé describes the bricolage-like perception of the traveler, who builds an impression from "movement of the landscapes which he catches only in partial glimpses, a series of 'snapshots' piled hurriedly into his memory and, literally, recomposed in the account he gives of them, the sequencing of slides in the commentary he imposes on his entourage when he returns" (1992/2008, p. 85). As impressions of these non-places are recomposed according to the logic of the tourist event, the LEGO City can be described as "a set of breaks and discontinuities in space that expresses continuity in time"

(p. 60). And as the self-reflexive LEGO Store winks from inside its representation, it serves as a reminder that this urban imagination is itself an ideal whose implied consumer is a suburban child.

Thus, even as this set brings the child to the city by simulating a day trip, it simultaneously brings the city to the child consumer. Situating LEGO's structural elements within the domestic space of play (freed from the regulating force of the playmat) creates a world that is an admixture of miniaturized fantasies and realities that represent a distant urban life embodied in LEGO bricks and sheltered by the comforting space of the carpet. Similarly, the train platform functions as a transitional space, a magic portal that connects toy play with urban fantasy. Essentially a frame, this platform is architecture open on both sides. Like the wardrobe opening into Narnia or Platform 9 ¾ in *Harry Potter*, the openness of this passageway—the quintessential non-place—is an invitation to fantasy play even while its realistic portrayal of public transit ties this fantasy to the mystique of urban life, whose 'gritty' reality is a source of fascination for suburban children.

Establishing an urban ethos governed by construction, protection, and transportation, LEGO City constructs a suburban fantasy of urban life as the fragmented stage of countless exciting urban dramas. This fantasy aims to appeal both to the nostalgia of parents who often purchase toys based on their idealized memories of childhood play and the present desires of children who demand stimulation in a media culture that allows direct marketing to children.[20] Thus, the urbanizing trend in LEGO construction play by no means replaces the suburban ethos of the Town Plan. Instead, this urban fantasy shelters a suburban ethos even as LEGO's indoor play shelters the playing child.

In LEGO, suburban and urban spaces are not only linked through transportation networks, they are presented as complementary, even entwined spaces. Similarly, one might say that modern life is characterized by a delicate balance between urban and suburban, public and private. As Bachelard puts it, "The two kinds of space, intimate space and exterior space, keep encouraging each other, as it were, in their growth" (1958/1994, p. 201). In this way, the differing messages of LEGO's houses, towns, and cities reinforce a surprisingly coherent domestic ideology. In the urban fantasy play of LEGO City, flirting with danger

[20] Gary Cross (2010) further explores how consumer culture encourages parents to take vicarious pleasure in children's consumption as *valves of adult desire*.

and activity from within the safe confines of the suburban space, work and commerce seemingly replace domesticity as the central reality of this interactive world. Yet, although LEGO City no longer directly represents domestic spaces, home is consistently implied as the protected space from which such imaginative forays can issue. Furthermore, a domestic ideal motivates this dramatization of urban life as construction and protection, ceaseless activities that maintain the sanctity of the domestic sphere despite being conducted outside it.

Construction play, therefore, links work and domesticity by emphasizing the interdependence of the public and private spheres, particularly by emphasizing the kinds of urban labor that construct and protect domestic spaces. Even in the absence of spatially represented domestic spaces, LEGO construction play is oriented around a suburban ethos that persistently domesticates its playful presentation of urban space.

Domesticating Space

Despite their quite different approaches, these LEGO houses, towns, and cities coalesce around a shared preoccupation with *domesticating space*, an ideological imperative that scripts construction play according to a domestic poetics of space. Scripting its play according to a decidedly suburban, domestic ethos, the LEGO medium promises to shelter the *development, imagination, creativity* (see Preface) of the playing child. LEGO is at once a *domestic* medium designed for quiet, indoor play, a *domesticating* medium designed to socialize children a domestic ideology of urban and suburban spaces, and a *domesticated* medium designed to easily snap into predetermined relationships. Thus, as the structural bricolage of construction play forms the foundation of all other forms of LEGO play, the ideology of construction extends well beyond explicitly architectural play themes. One might say that domesticating space constitutes the structuring logic of this structural medium and the architecture of this architectural toy.

In construction play, domestication is scripted not only as a sheltering ideal but also as a design imperative. In other words, the material relationship that LEGO cultivates between player and toy medium is particularly prone to a fantasy of mastery. As I argue elsewhere (2019), mastery is one of the core ideologies cultivated by the LEGO company, whose rhetoric of expertise is commonly described as “master building.” With respect to the domesticating spaces of construction play, the fantasy of mastery inflects the aforementioned ideological construction of LEGO's miniature houses, towns, and cities. And, in so doing, it constructs an ideology of the playing child as using toys to bridge the nostalgic and technological.

An ideology of mastery redefines the significance of the miniature by providing the player of miniature toys with more agency and control over the experience of a miniature sublime that depends upon the interplay between accessibility and inaccessibility. The accessibility of hands-on play transforms miniature fantasies otherwise mediated only by the gaze. Thus, Susan Stewart describes the mediated fantasy of the miniature toy, writing that:

> The toy is the physical embodiment of the fiction: it is a device for fantasy, a point of beginning for narrative. The toy opens an interior world, lending itself to fantasy and privacy in a way that the abstract space, the playground, of social play does not. To toy with something is to manipulate it, to try it out within sets of contexts. (1984, p. 56)

Offering a tangible, manipulable world, toys invite players to directly interact with miniature fantasies. Toys lend their material concreteness to the sublime in-between of the miniature—their tangible play mediating the constitutive tensions of conceptual interplays.

Materially, miniature toy culture relies upon rather than protects domestic space. Ideologically, however, the easy manipulation of the miniature toy allows such toys to perform a fantasy of mastery over the miniature world. Consequently, miniature toys are *mediums*, not in the modern sense of being communications media, but in the older, spiritualism-inflected sense of mediating between the material and immaterial—or, in this case, between physical object-relations and psychical longing.[21] Perhaps, the allure of both spiritual and communications mediums rests in a fantasy of mastery, a deep-seated desire to control the flow of meanings that circulate around us (after all, participatory media culture has managed not only to commodify media texts but also *access* to media texts). In this spirit, miniature toys open up evocative yet inaccessible miniature worlds to tangible control.

LEGO takes this a step further, not only allowing players to manipulate the miniature world, but to *construct* it. This transforms the miniature

[21] See Karen Beckman's (2003) *Vanishing Women*, particularly a discussion of the spirit medium's gendered role in Chapter 2.

experience from a matter of inhabitation to a matter of creation, inverting the core dependency between player and world. LEGO construction play is *creative* in part because this docile toy medium constitutes a world in which creation and re-creation are extremely accessible. Playing out a fantasy of modern plasticity (Lee 2014), these toys play out the boundless potentiality of a single malleable substance that can take on any imaginable form. Indeed, a central joy of material LEGO is the very tangible and somewhat hubristic pleasure of toying with matter that facilitates its own construction. LEGO, that is, manages to exemplify the int(e)ractability of matter in a way that feels more interact-able and less intractable.

The miniature suburbs and cities of the Town Plan and LEGO City, moreover, inflect this core fantasy of mastery with a more particular ideological construction than the general anthropocentric privileging of humankind over the material world. In these social constructions, the fantasy of mastery is aligned with the establishment of a middle-class administrative class exerting regulatory agency over the process of construction. For instance, the Town Plan and LEGO City sets typically situate the player as performing white-collar intellectual labor—mayor rather than citizen, or architect rather than construction worker. The minifigure characters included in these sets often leave the absent administrative roles open for the player to role-play while providing tangible representations of biopower (i.e. citizens and laborers) for the player to manipulate. Consequently, LEGO construction play becomes associated more with the abstracted labor of conceptualizing a building than the more tangible labor of constructing it. Rather than rendering LEGO non-ideological, this abstraction reaffirms LEGO's ideological commitment to a cultural ideal of technological mastery in which the pseudo-objective sciences of architecture and engineering regulate the material world. Consequently, LEGO's construction themes implicitly construct this creative construction toy as a way of materially playing out the domestication of space.

LEGO construction play, therefore, exhibits a tension between a tendency to socialize children and domesticate their play and a contradictory tendency to unleash children's potential for creative mastery over the medium. This is perhaps unsurprising, as childhood is already conflicted culturally-constructed space that synthesizes nostalgic (childhood as a space of innocence and purity) and progressive (childhood the unrealized promise of futurity) impulses. Anchoring this nostalgic ideal of childhood is an association with naturalness. While this association often also connotes wildness—consider how the image of the child as a *noble savage*[22] untainted and untrained by civilizing forces—the ideology of LEGO construction play aligns the natural more with domesticity. Consider how the LEGO Creator Treehouse weaves its natural world into its domestic playspaces,[23] how the Town Plan exemplifies a suburban ethos by presenting only cultivated, contained nature, and how LEGO City tames its outskirts (several of its wilderness-themed sets are oriented around logging or forest police, subsuming nature under the aforementioned logics of construction and protection). In the regulated LEGO world, nature never overruns technological mastery, but instead becomes a part of its regulatory system. Similarly, within this domesticating ideology, the nostalgic figure of the 'natural' child is positioned as simultaneously the target and agent of domesticating LEGO play.

The deeply regulatory impulse that underlies LEGO construction play is significant *because*—not in spite of—the fact that LEGO is fundamentally ludic, playful, and fun. Indeed, the playfulness associated with childhood is a large part of the imaginary of the 'natural' child. And this association contributes to the nostalgic construction of LEGO play. As artist Nathan Sawaya notes, "The nostalgia of LEGO bricks weighs heavily on the viewer of LEGO art" (2014, p. 212). And, as Grosz points out, echoing Bachelard's notion of *felicitous space*, "There is a certain joy in our immersion in space. It is important to recognize that you can attain a certain (temporary) depersonalization and still enjoy it, enjoy the expansion and permeability of bodily boundaries" (2001, pp. 20–21). Indeed, part of the pleasure of this complex architectural toy is how it plays out profound contradictions—between inside and outside, suburb and city, security and danger, construction and deconstruction—while retaining a nostalgically domestic, domesticated, and domesticating aura.

To play with LEGO is to play at construction—material, ideological, social. Materially, this entails the tangible bricolage of piecing

[22] According to Barbro Johansson, "The effect of this 'othering' process is that children are marginalized and sometimes also romanticized in the same way as women and the 'noble savage' were attributed higher moral and more sublime characteristics than the common man" (2010, pp. 80–81).

[23] This is enhanced by the ontological continuity between 'natural' and 'technological' LEGO, both of which are materially embodied in a single plastic substance. Interestingly, LEGO has begun to literally infuse the natural world into its plasticity by experimenting with sustainable plant-based plastics.

together already-significant elements. Ideologically, this entails invoking and playing with already-significant cultural constructions of childhood, construction, and play. Socially, this entails domesticating play according to the regulatory ideals of the suburban ethos. To play at construction is to play with and play out architectural formations—material, ideological, social. Inspired by Bachelard, I opened this chapter by speculating that *our selves are built environments, and by playing at houses and rooms, we learn to reconstruct ourselves*. If we take this possibility seriously, construction play becomes a *performative* exercise in constructing and reconstructing identity, an inextricable complement to the subject of the next chapter—*dramatic play*.

Post-script 2—Mike Doyle's Deconstruction Play

Built and photographed by the author of *Beautiful LEGO*, Mike Doyle, *Victorian on Mud Heap* (Fig. 2.7) is the third in a series of haunting images of black-and-white LEGO houses decaying as they collide and merge with 'natural' elements such as trees, snow, and mud rendered in LEGO. These images of decay deconstruct the utopic presentation of domesticating space that characterizes miniature LEGO—first by using subversive building techniques that disrupt the regularity of LEGO construction and second by destabilizing the regulatory power technologized built environments exert over nature, as natural forces reclaim their wild and unruly spaces.

Describing this build as a "textural exploration of decay" (2011b, n.p.) in his Flickr artist statement, Doyle deconstructs the architectural utopianism of miniature LEGO by disrupting the formal regularity of standard LEGO construction. Although all the elements of his build are standard, unmodified LEGO elements, Doyle uses a variety of subversive building techniques that reimagine the fundamental design principles of LEGO construction (as many fan creations do). For instance, whereas standard LEGO construction adheres to the Cartesian grid built into LEGO design, Doyle takes advantage of the degree of play present in all LEGO connections to disturb this regularity. By slanting various lines of the structure by a few degrees and introducing gaps and deviations within repeated patterns, Doyle disrupts the visual harmony of the architectural design. Indeed, the slightness of many of these deviations only intensifies the feeling of the uncanny valley by making it easy to imagine the undecayed structure. Thus, these techniques remain solidly within

Fig. 2.7 Mike Doyle's *Victorian on Mud Heap.* Although this six-foot-wide structure conveys a feeling of decay, no LEGO elements were altered or distressed, and no non-LEGO elements were added. Also, the black-and-white appearance is due entirely to the color of the bricks used (Image courtesy of the artist)

the boundaries of what LEGO *can* do while challenging the implicit design imperatives to pursue regularity, the scripts that reveal what LEGO elements are *designed to* do.

Whereas his Victorian house primarily plays on slight, uncanny deviations from the standard rigidity of LEGO construction, Doyle's mud heap relies on more unique and clever construction techniques. The mud heap consists of a substructure of LEGO hoses, whose flexibility provides "huge control over the angle of the plate in three dimensional space which contributed greatly to on organic flow" (2011a, n.p.). The jumbled elements that make up the heap are then draped over this organic substructure. Interestingly, even as the design of this substructure is quite unlike standard LEGO construction, it has its own architectural (rather than natural or mud-like) character. Attaching surface elements to a skeletal substructure—even an organically-shaped one—is a viable architectural technique reminiscent of modernist *curtain wall* construction, a technique Tom Alphin (2015) details in his LEGO guide to architectural design. At the same time, the flexible, sculptural substructure subverts the domesticating regularity of typical Cartesian LEGO architecture, fusing the architectural and the natural.

Thus, Doyle's artistic bricolage deconstructs the utopic messages of LEGO without subverting the architectural foundations of the medium itself. Indeed, it might be more accurate to say that Doyle plays with the hidden possibilities of the medium, revealing an organic, sculptural potential implicit in LEGO design. In so doing, Doyle's build challenges the dominant conception of LEGO as strictly Cartesian and introduces tensions between nature and artifice as being constitutive of the medium.

Much of the rhetoric surrounding the development of the modern suburb involved harmoniously blending city and country, creating a domestic environment that provided both modern conveniences and picturesque, cultivated nature. In Doyle's triptych, however, nature threatens architecture with *decay*, the incursion of wild or unruly nature as a destructive/deconstructive force against domestic architecture—the very forces houses are designed to shelter their inhabitants from. As the natural elements of mud, trees, and snow intrude into the sanctity of the house, its interior becomes open to the environment, literally eroding the distinction between inside and outside. To be a house is at some level to exclude nature. Thus, as nature enters these abandoned houses, they transform from home to relic—a reminder that intensifies the sense of loss behind miniature representations.

The tension between the naturalness of decay and the artificiality of architecture is heightened by the juxtaposition between the elements of design and nature that give the build its name: *Victorian on Mud Heap*. These elements are visually oriented as a vertical continuum—indicated by the 'on' in the title—from the depths of the mud to the heights of the house's spires. Like archaeological strata, these layers spatialize epochal time as the natural force of the mud lies in the depths, eroding the architecture above even as the architecture disintegrates back into these depths. The significance of this build, therefore, depends on the tension between these two poles, each of which depends upon a metaphorical reading of extant architectural symbols. At the uppermost point of this build, which exhibits the least visible decay, Doyle mimics the utopian character of existing architectural symbols in a build that "echoes the gothic representation of cathedrals – with their many spires – reaching upward to the heavens" (2011b, n.p.). At its lowermost point, however, Doyle deconstructs architectural utopianism by burying the house's foundation under the sloughing chaos of the mud heap. Further challenging the metaphorical connotations of the foundation, Doyle writes, "Foundations give way. Permanence transmutes into fragility. Our safe havens betray us" (2011b, n.p.). As the gravity of this build pulls architecture into its deconstructive depths, utopian domestic architecture dissolves into a study in decay.

In so doing, this deconstruction play challenges not only the utopic character of architecture but also the utopic character of LEGO play as a domesticating playspace. Beyond deconstructing modern suburbia, this art also deconstructs the nostalgic ideal of the Victorian dollhouse as a domestic space for sheltering play. The violent meeting of architecture and decay, therefore, contrasts with the decontextualized ideal of the indoor toy. This decontextualization is heightened by how this structure is framed: foregrounded against a backdrop of empty space filled only with an eerie play of light and dark. This ethereal presentation contextualizes the house not within the physical world but within the subconscious, like a dollhouse that is decontextualized from its surroundings and recontextualized within the psychic space of imaginative play.

Accentuating the feeling of an eerie subconscious depth, the mud can be read not as encroaching from the outside but as emerging from the inside. As "mud travels through the first floor, tears down a front wall and oozes over the porch side, taking with it household contents of convenience" (2011b, n.p.), Doyle's house becomes a passageway for a primordial flow that precludes inhabitation. Recalling Bachelard's suggestion that cellars represent the subconscious, this subterranean curtain of mud subverts an idealized picture of dollhouse play by overflowing its depths, thereby exposing the architectural veneer as merely concealing the darker and deeper aspects of the psyche.[24]

[24] As it happens, studies on actual doll play reveal this idealized picture to not be quite historically accurate. In fact, Miriam Formanek-Brunell (1993), Robin Bernstein (2011), and others have noted a long tradition of darker forms of doll play, such as physically mutilating or abusing dolls or performing somber doll funerals.

Another quite different layer of deconstruction play can be found in the tension between the medium of LEGO and that of photography, with which this build is captured. While online fan communities regularly display constructions through photography, Doyle's build is a façade specifically designed to be remediated in two-dimensional photography.[25] Ironically, Doyle uses the flattening medium of photography to "play on depth," as he describes in an online interview:

> In this series, I am most interested in textures and the effect of layering textures over each other. To this end, the absence of color helps the viewer to focus on just this. Lego colors tend to be pretty harsh and unrealistic for my tastes, so I stick to black/white and grays. **Without color, we dive right into form, which is where I want you to be**. (qtd. in Thita 2011)

Desaturating LEGO colors to emphasize the play of depth and texture of a paradoxically flattened image, Doyle emphasizes the expressive potential of the medium by stripping away much of its media-specific character. Without suggesting that LEGO is primarily a visual medium, Doyle draws upon visual plays of light and dark to strikingly convey an *idea* of his structure's absent three-dimensionality. Like the miniature sublime, this photographic remediation makes LEGO feel tangible while denying the touch of the viewer.

As Doyle's build uses LEGO to remediate dollhouse architecture and photography to remediate LEGO, it reframes LEGO as something other—and thereby something more—than a utopic construction medium. That is not to say, however, that Doyle's architecture is dystopic. Unlike an inherently repressive or afflicting dystopic space, this build represents a utopic domestic space in the process of decay, a slow dissolution back into nature. Consequently, the resultant tone evokes a kind of inverted or bittersweet nostalgia. Like everyday utopic nostalgia, this build remembers the idyll of a bygone past. Unlike this nostalgia, however, this build aestheticizes the process of decay rather than the lost idyll. In so doing, it evokes wistful, nostalgic cultural memories in order to deconstruct the nostalgic ideal, performing a kind of sublime deconstruction play.

[25] In contrast, many LEGO constructions posted online include photographs taken from various angles to recreate the sense of three-dimensionality that can be lost when built structures are captured in photography.

Works Cited

Alphin, Tom. 2015. *The LEGO® architect*. San Francisco: No Starch Press.

Anderson, Benedict. 2000. *Imagined communities: Reflections on the origin and spread of nationalism*. New York: Verso.

Augé, Marc. 2008. *Non-places: An introduction to supermodernity*. Trans. John Howe. New York: Verso.

Bachelard, Gaston. 1994. *The poetics of space*. Trans. Maria Jolas. Boston: Beacon Press.

Baichtal, John, and Joe Meno. 2011. *The cult of LEGO*. San Francisco: No Starch Press.

Beckman, Karen. 2003. *Vanishing women: Magic, film, and feminism*. Durham: Duke University Press.

Bernstein, Robin. 2011. *Racial innocence: Performing childhood and race from slavery to civil rights*. New York: New York University Press.

Bogost, Ian. 2006. *Unit operations: An approach to videogame criticism*. Cambridge: MIT Press.

Cross, Gary. 2010. Valves of adult desire: The regulation and incitement of children's consumption. In *Childhood and consumer culture*, ed. David Buckingham and Vebjørg Tingstad, 17–30. London: Palgrave Macmillan.

Doyle, Mike. 2011a. The making of... part 1. *Snap*. http://mikedoylesnap.blogspot.com/2011/09/making-of-part-1.html. Accessed on 25 June 2018.

Doyle, Mike. 2011b. Victorian on mud heap. Flickr. https://www.flickr.com/photos/7931559@N08/6192835432/. Accessed 15 April 2020.

Edelman, Lee. 2004. *No future: Queer theory and the death drive*. Durham: Duke University Press.

Formanek-Brunell, Miriam. 1993. *Made to play house: Dolls and the commercialization of American girlhood, 1839–1930*. New Haven: Yale University Press.

Foucault, Michel. 1995. *Discipline and punish: The birth of the prison*. Trans. Alan Sheridan. New York: Vintage Books.

Freeman, Elizabeth. 2010. *Time binds: Queer temporalities, queer histories*. Durham: Duke University Press.

Giddings, Seth. 2014. Bright bricks, dark play: On the impossibility of studying LEGO. In *LEGO studies*, ed. Mark J.P. Wolf, 241–267. New York: Routledge.

Grosz, Elizabeth. 2001. *Architecture from the outside: Essays on virtual and real space*. Cambridge: MIT Press.

Johansson, Barbro. 2010. Subjectivities of the child consumer: Beings and becomings. In *Childhood and consumer culture*, ed. David Buckingham and Vebjørg Tingstad, 80–93. London: Palgrave Macmillan.

Kullman, Kim. 2009. Enacting traffic spaces. *Space and Culture* 12 (2): 205–217.

Lauwaert, Maaike. 2009. *The place of play: Toys and digital cultures*. Amsterdam: Amsterdam University Press.

Lee, Jonathan Rey. 2014. The plastic art of LEGO: An essay into material culture. In *Design, mediation, and the posthuman*, ed. Dennis M. Weiss, Amy D. Propen, and Colby Emmerson Reid, 95–112. Lanham: Lexington Books.

Lee, Jonathan Rey. 2019. Master building and creative vision in *The LEGO Movie*. In *Cultural studies of LEGO*, ed. Rebecca C. Hains and Sharon R. Mazzarella, 149–173. Cham: Palgrave Macmillan.

LEGO Architecture Studio [booklet]. 2013. Set 21050.

Lipkowitz, Daniel. 2012. *The LEGO book*, 2nd ed. London: DK.

McCloud, Scott. 1994. *Understanding comics: The invisible art*. New York: HarperPerennial.

Millhauser, Steven. 1983. The fascination of the miniature. *Grand Street* 2 (4): 128–135.

Ogata, Amy F. 2013. *Designing the creative child: Playthings and places in midcentury America*. Minneapolis: University of Minnesota Press.

Robertson, David C. 2013. *Brick by brick: How LEGO rewrote the rules of innovation and conquered the global toy industry*. New York: Crown Business.

Sawaya, Nathan. 2014. LEGO: The imperfect art tool. In *LEGO studies*, ed. Mark J.P. Wolf, 206–215. New York: Routledge.

Stewart, Susan. 1984. *On longing: Narratives of the miniature, the gigantic, the souvenir, the collection*. Baltimore: Johns Hopkins University Press.

Sutton-Smith, Brian. 1986. *Toys as culture*. New York: Gardner.

Thita. 2011. Featured LEGO artist: Mike Doyle. *The Brick Blogger*. http://thebrickblogger.com/2011/09/mike-doyle-lego-artist/. Accessed 15 April 2020.

Wolf, Mark J.P. 2014. Adapting the Death Star into LEGO: The case of LEGO set #10188. In *LEGO studies*, ed. Mark J.P. Wolf, 15–39. New York: Routledge.

CHAPTER 3

Playing House with LEGO Friends

As architectural design anticipates inhabitation, housing play anticipates playing house.[1] Yet, these otherwise complementary modes of play evoke quite different cultural ideologies. Housing play evokes a masculinized ideal of the rational organization of society that can be glimpsed in the Town Plan's suburban planning and LEGO City's construction equipment, transit systems, police and fire themes, and so on. Conversely, playing house evokes a feminized domestic labor exemplified by the 2012 'for girls' product line LEGO Friends' main themes of social activities, animal care, and food service. This unsubtle thematic disparity reflects familiar cultural codes for gendered toys, codes which one recent study found that its girl subjects had no trouble decoding, as they "could easily describe for us the characteristics of LEGO toys that indicated whether or not the company intended them for girls or for boys" (Hains and Shewmaker 2019, pp. 253–254). This is, of course, intentional, as LEGO Friends "was made with the goal of inspiring more girls than we currently serve to try their hand at building and experience the pride of accomplishment that LEGO play fosters" (Nipper 2012, n.p.). In other words, the explicit rationale for thematically gendering LEGO play is to invite girls into construction play, a superficially laudable goal that unfortunately masks a set of deeper ideological assumptions in which construction play

[1] Some LEGO construction sets are clearly oriented towards *playing house* in a very traditional sense. For instance, the Homemaker line from the 1970s features very dollhouse-like furniture and nuclear family figures that predate the modern minifigure.

J. R. Lee, *Deconstructing LEGO*,
https://doi.org/10.1007/978-3-030-53665-7_3

is itself gendered masculine while the dramatic play involved in playing house is gendered feminine.

To contextualize this analysis, it is worth briefly exploring critiques levied at LEGO by activists such as Anita Sarkeesian and the SPARK Movement in the wake of the LEGO Friends announcement. One important thread of this critique is that explicitly designing a girls' version of LEGO might not have been necessary if the history of LEGO had not already demonstrated a steady drift from relative gender-neutrality towards boy culture. Lori Landay reports that "LEGO has its own internal mythologizing of types of boys that it uses in the development process" (2014, p. 74), referring a set of four named male archetypes used by LEGO to envision their consumer base. And this gendered imagining is reinforced by advertising figures such as Zack the LEGO Maniac and LEGO Club spokes-figure Max, who both lend their youthful masculinity to the portrayal of the 'cool' LEGO consumer. Unlike the celebration of childlike creativity in the relatively gender-neutral advertisement (Fig. 3.1) starring Rachel Giordano, these figures advertise LEGO by performing desirable gendered, an advertising strategy based entirely on aspirational identities rather than the affordances of the product. This LEGO is 'for boys' because it situates LEGO play within a masculinized play culture in which toys serve as gendered status symbols.

Within this gendered play culture, the design of LEGO products also reflected a masculine bias prior to LEGO Friends, exhibiting a massively imbalanced ratio of male to female minifigures[2] and a marked rise in violent play themes (Bartneck et al. 2016). Thus, when LEGO neutrally noted that "our active household studies indicate that we have not been as successful in drawing the interest of more girls with what we currently offer" (Nipper 2012, n.p.), it seemed willfully incognizant of the fact that its current offerings were gendered masculine rather than gender-neutral. It is tempting here to read this masculinization of LEGO as a layer of thematic signifiers atop an otherwise gender-neutral medium. Indeed, many consumer activists used nostalgic language in calling LEGO to return to its more abstract roots, for instance, citing a founding LEGO principle that stated that LEGO should be "For girls and for boys"

[2] One often-cited figure is a ratio of 18-to-1 (Baichtal and Meno 2011, p. 57). And even with a count that admittedly gives LEGO the benefit of the doubt in all dubious cases, David Pickett's (2012) year-by-year counts from 1989 to 2011 range anywhere from about 13-to-1 to 2-to-1.

Fig. 3.1 A 1981 magazine advertisement for Universal Building Sets featuring a young Rachel Giordano (see Preface). While LEGO toys and advertising were never wholly disconnected from gendered ideologies, this advertisement neither excessively genders Giordano nor her construction. This more gender-neutral play is typically targeted at younger children, as the use of Duplo in Giordano's creation implies

(Baichtal and Meno 2011, p. 14) and a magazine advertisement depicting a young girl proudly displaying her abstract construction (see Preface). And there is certainly some truth to this, as the gendering of LEGO rather noticeably increases over time.

Yet, at the same time, beneath the problematic decision to inflect its construction play with masculine play themes, lies a deeper issue—that LEGO seems to have gendered even the most abstract forms of construction play as masculine. Indeed, Jørgen Knudstorp, the LEGO CEO who oversaw the release of the Friends line, is on record as saying, "There is something about the idea of constructing and deconstructing or destroying which frankly is an important part of Lego play that is a very boys-type of activity" (Bender 2010, p. 81). In making this problematic statement, Knudstorp echoes a long line of thought that has historically gendered both STEM education and the role of construction toys in furthering such education. For instance, his rhetorical paralleling of construction and destruction resembles how "Advertisements of Erector Set described boys' nature as both constructive and destructive" (Zinguer 2015, p. 116). As Zinguer further explores, both Erector Sets and Meccano cultivated a culture of boys play inspired by Samuel Smiles *Self-Help*, which reframed the history of innovation as the tangible result of a kind of masculinized spirit of creative independence and dogged perseverance. To facilitate this, "both toy publications instilled the sense of "shop culture" that playing with Erector and Meccano provided" (p. 116). While LEGO was always more a partially abstracted architectural toy than a pure engineering toy like Erector and Meccano, Knudstorp's comments hint that the gendering of LEGO play may implicitly reflect a more widespread cultural gendering of play.

In this context, the intensive "four years of research in play needs of girls" that LEGO underwent to develop "the first 100 percent LEGO building experience fully optimized to girls' tastes and interests" (Nipper 2012, n.p.) is a move to inclusivity grounded in a highly gendered set of assumptions. Rather than assuming that girls are interested in building—as LEGO assumes boys naturally are and the girls in Hains and Shewmaker's study seem to already be—LEGO assumes a natural disconnect between girls and construction play that needs to be 'solved.' As Knudstorp explains, "We felt it was time to test assumptions that girls aren't interested in building" and to do so "We focused on creating a play experience centered on the joy of creation, while heeding the way girls naturally build and play" (Nipper 2012, n.p.). Even in acknowledging that girls *can* be interested in construction play, Knudstorp suggests that this interest in contingent on construction play being suitably aligned with more 'natural' forms of girls' play, one which the press release describes as characterized by "more details and interior building, a brighter color palette, a more realistic figure, role play opportunities and a story line that they would find interesting." (Nipper 2012, n.p.). While their market research may have shown that many girls do prefer these things (as I am not privy to this data, I cannot verify these conclusions), it is highly doubtful that any LEGO research has shown these correlations to be 'natural' rather than culturally-constructed.

And this cultural construction is highly problematic. First, the type of play that underlies LEGO Friends aestheticizes design, orienting construction play around the feminized labor of interior design rather than the more masculinized LEGO themes of engineering, construction, architecture, and suburban planning. This suggests that while girls might have equal share in construction play, their share is in a fundamentally different and thoroughly domestic type. Second, this reframing of construction play also comes with a decreased investment in construction play to instead emphasize *dramatic play*, playing house within LEGO playsets. This is especially significant ideologically, as the theatrical play (play as performance) that defines dramatic play is inextricably linked to identity play (play as performative).

Characteristically, this chapter is set against a backdrop of mixed messages from LEGO, which belies how construction play and dramatic play are necessarily intertwined by increasingly gendering these forms of play. This introduces a gendered schism not only into individual products but also into the medium itself and even into the modes of play it engenders. Tracing the gendering of dramatic play from early thematic sets to LEGO Friends, this chapter deconstructs the theatrical staging of LEGO set design, traces the socializing messages of such stagings, performs a comparative analysis of the representation of gender in LEGO minifigures and LEGO Friends mini-dolls, and finally reflects on the consequences of introducing a gendered division within a single construction medium.

Setting the Stage

The central feature that defines LEGO sets—the most common form in which LEGO is marketed and sold—is an invitation to drama. As play unfolds through an intimate negotiation with the space it occupies, toy architecture explicitly sets the stage for implied dramatic play. For the most part,[3] most LEGO sets are also *playsets* that are explicitly designed to provide a material and conceptual architecture that literally sets the stage for dramatic play While for many toys (for instance, Barbie), playsets are marketed as accessories, LEGO inverts the conventional relationship between doll and dollhouse by giving playsets primacy,[4] allowing LEGO to maintain a core identity as a spatial construction toy while increasingly emphasizing dramatic play with LEGO minifigures. Comparing LEGO to the traditions of dollhouses and toy theaters, this section examines how the design of LEGO playsets sets the stage for dramatic play by framing play as performance.

Although dolls are some of the earliest known toys, the dollhouse is a more recent cultural development, deriving from a miniature tradition (see Chapter 2) of upscale collectibles that evolved into the now-familiar spaces for playing house. While not the norm, LEGO offers a few sets explicitly designed for playing house, such as City House (Set #8403) and several houses from the LEGO Creator and LEGO Friends lines (see Chapter 2). Unlike traditional dollhouses, however, these sets tend to have cramped interiors and place more emphasis on windows, doors, and exterior spaces, suggesting that they are still somewhat beholden to the suburban sprawl of the Town Plan (see Chapter 2).

Where LEGO playsets most closely reflect dollhouse design is in their architecture. Since the hermetically-sealed structures of the early LEGO Town Plan, most LEGO structures allow some manner of player access. This can be accomplished through roofs that lift off, walls that hinge open, or—most commonly—an absent fourth wall. Like a sitcom set or dollhouse, an absent fourth wall invites participation in an otherwise enclosed and typically domestic space (sitcom viewers are invited as voyeurs and dollhouse players are invited as puppeteers). For toys, this design not only allows players to intervene within the toy world but forges material and symbolic continuity between the toy world and child's playspace. Indeed, the absent fourth wall is what allows the other three walls to shelter, bound, and thereby domesticate the playspace. This spatial orientation aligns the physical performance of dramatic play with

[3] Perhaps the most notable exception is the LEGO Architecture line, which emphasizes display over play.

[4] LEGO has produced more traditional dolls and playsets, such as the second iteration of Scala (1997–2001).

the internal imagination of the child. As Frances Armstrong (1996) notes of dollhouse play, while such private spaces are literally and figuratively confining, their privacy also harbors subversive potential by freeing the dramatic imagination from surveillance.[5]

Although this dollhouse dynamic is certainly at play in LEGO, many playsets diverge from this tradition by offering a more public orientation. One ubiquitous design style, of which the Pizza To Go set (Fig. 3.2) is a representative example, features an absent fourth wall, a structure with relatively little interior depth, and a partially absent second wall that allows the façade to frame the interior play. While opening the fourth wall may facilitate surveillance, it primarily functions to orient the playset around *performance*, framing the architectural space in a way that implies spectatorship even if no spectator is present.

Such designs situate LEGO playsets within another toy tradition, namely the miniature stages of the Victorian toy theater. Like LEGO, toy theaters derived a significant part of their appeal from the process of their construction, as toy theater historian George Speaight argues: "The great thing about the real toy theatre was the time and labour it occupied, the painting, the cutting out, and the preparations" (1969, p. 145). And like toy theaters, LEGO derives a significant part of its appeal from its orientation as a construction toy designed to facilitate dramatic play. In other words, despite the LEGO medium being primarily advertised for its construction play, the teleology of most LEGO products as designed points toward an ongoing dramatic play enacted upon the miniature stages of completed playsets.

As is typically the case with LEGO, this invitation advertises the specificity of the medium as a means of drawing out the creative potential of the player. In other words, just as the LEGO brand advertises its construction toys as promoting creativity it advertises its thematic playsets as promoting imagination, recalling the trinity of *development, imagination, creativity* discussed in the Preface. One particularly illuminating advertisement (Fig. 3.3) uses a theatrical stage to metaphorically link

[5] Armstrong writes on the one hand, "A problem of demarcated ludic spaces is that a child given a special place to play will often be subject to authority in that area: even if not confined to it as to a prison, she may be watched while she plays there. The panopticonlike structure of a dollhouse also lays the dolls themselves open to surveillance" (1996, p. 27). On the other hand, she notes that "because a dollhouse offers accessibility and privacy, even a heavily supervised child may commit acts of innovation or transgression while ostensibly dusting and tidying" (p. 35).

Fig. 3.2 Product art for the 1994 Pizza To Go playset (Set #6350, rereleased in 2002 as #10036). All three characters are posed in mid-action, clearly showing their respective roles within this ritualized restaurant play. The large counter window is shown to be a transitional space in two ways: (1) the two pizzas on the counter with no narrative context enhance the liminality of the space and (2) although this image does not show pizzas passing through this window, the way the chef and customer flank the window suggest that the depicted transaction depends on bridging this space

construction and dramatic play, as a young girl builds LEGO structures as a backdrop for an amateur theatrical performance, she bows in front of her creations, revealing that her creative expression—not the toys themselves—is the true performance. This advertisement features a strong developmental narrative as the girl narrates a persuasive (i.e. public) argument defending her mode of self-directed play. Implicitly addressing her mother, she asserts her independence ("I want to figure it out on my own") while simultaneously admitting her dependence ("because you taught me how to think") before synthesizing the two in her final appeal ("I'm about to make something that I know will make you proud").

Fig. 3.3 Screenshot from the 2014 LEGO advertisement "Inspire Imagination and Keep Building" on YouTube. This image shows several layers of theatrical framing: the two shadowed heads of the audience (presumably parents), the curtained stage peppered with lights, and the LEGO backdrop lit up in reverse projection. The girl is depicted in mid bow—not performing any particular drama but performing theatricality itself

In this moment, play itself becomes a performance of identity, an identity shaped by the intertwined developmental narratives of creativity and imagination, construction and dramatic play. As is common in advertising children's products in general and LEGO in particular (Lee 2019), this suggests to parents that their children's self-actualization hinges on the parents' ability to provide a protected and supportive developmental environment. This is an invitation not only to consume LEGO products but also to encourage a uniquely LEGO brand of creativity and imagination, performed via bricolages of material and symbolic signifiers—the systematic assembly of LEGO elements into architectural stages and the subsequent and scripted role-play enacted upon such stages.

This invitation to the more dramatic elements of performative play fundamentally alters and, in some sense, completes the relationship between player and medium established in construction play. Unlike both the mythical promise of a player expressing creativity through abstract bricks and the more material promise of a player surveying a miniature suburban development from the administrative perspective of planner and regulator, these miniature stages do not situate players as creators of miniature worlds. Players are invited to enter into and act within, not merely upon, these worlds.

Although the dramas that unfold are not literally scripted (the way toy theaters often provided condensed versions of contemporary theatrical scripts), LEGO playsets are designed to stage highly representational and ideologically-laden dramas starring highly representational and ideologically-laden toy characters. It would be tempting here to argue that this is problematic because representational toys overdetermine children's naturally more free and creative storytelling impulses. Yet, without a single corporately-designed prop, children can and do play out extremely scripted narratives when playing house. Given that children play as they are attempting to make sense of the cultural contexts that shelter and encourage such play, it is somewhat questionable what truly 'free' play might be. The more pertinent issue, to which we shall now turn, is *how* the specific designs of LEGO playsets and figures set the stage for performing various ideological narratives, especially—in this case—gender.

Set Design: Staging Gender

A stage invites performance merely by existing and thereby offering an architectural space that affords performative engagements between players and spectators. Toys invite play similarly. In a serendipitous linguistic affinity, *play* names both theatrical and toy performances, suggesting that these cultural rituals are intertwined even beyond the toy theater. Similarly, LEGO set design functions like theatrical set design in that the textual formation of LEGO playsets architecturally scripts certain interactions. Set design layers more specific meanings and possibilities onto the stage, physically orienting it into a space for enacting very particular types of drama. This happens in two primary ways. First, set design thematizes the stage, evoking a contentful setting that becomes the place of theatrical or toy play. For LEGO, this mostly means that its playsets thematically reinforce fairly straightforward narratives, such as police sets implying "cops and robbers" play. Second, set design provides architectural affordances that may script narrative events. For LEGO, this means that playsets offer features ranging from the relatively simple faucet and pizza oven in Pizza To Go to police holding cells with clever breakaway walls that incline players to stage jailbreaks. These details weave specific settings into the general staging, offering a bricolage of social signifiers that both reflect and promote stereotypical role-play.

Reflecting modernity's move toward more representational socializing toys, earlier thematic sets like Pizza To Go typically deploy bricolages of highly stereotypical cultural signifiers to frame their role-play. Such sets are highly generic, representing not specific places but rather the *ideas* of such places—like toy theaters, such set design accentuates significant details at the expense of realistic representation. For instance, the pizza oven and sales counter account for the entirety of the pizzeria's interior space. These details elicit role-play as the space itself suggests certain activities (baking and selling pizza), activities paratextually reinforced by box art showing the minifigure characters posed in the midst of such activities.

For such material playscripts, representation entails socialization. According to French theorist Roland Barthes, toys which are "essentially a microcosm of the adult world" must "*always mean something*, and this something is always entirely socialized" (1957/1972, p. 53). Like the domestic role-play of playing house, this mode of dramatic play entails a scripted playing out of archetypal social interactions. Lamenting that such toys have replaced traditional building blocks, Barthes draws a clear division between two developmental narratives, with the abstract cognitive benefits of construction play being valued above leveraging children's play for socialization.

This is precisely the line of critique aimed at LEGO Friends by activists who argued that these toys were too light on construction and too heavy on stereotypical social messages. Although these concerns are legitimate, there are problematic consequences of bifurcating these developmental narratives in stark binary terms. In addition to presupposing a fundamental incompatibility between construction play and dramatic play, this binarization may perpetuate the belief that the former is superior to the latter, a view subtly influenced by a Western history of valuing culturally masculinized rationality (an imposition of order aligned with construction play) over culturally feminized sociality (an exploration of feeling aligned with dramatic play). Care must be taken, therefore, to ensure that critiques of socializing toys deconstruct not only the specific messages encoded in the toy representation but also the larger ideologies that inflect and sometimes gender performative play itself.

The primary ethical question here is, therefore, less *whether* such toys are socializing and more *how* such toys socialize. After all, the educational toys (see Preface) and construction toys (see Chapter 2) that are so commonly held up as the ideal in these critiques are just as socializing as any other toy. To deconstruct the ideological construction of specific socializing toy designs, therefore, one might consider how a playset aligns stereotyped social identities with archetypal social roles.

In Pizza To Go, the characters function as agents of a socializing play based on the rules of retail exchange promised by the counter around which this set is architecturally oriented. Opening the second wall to create the theatrical frame, the counter also creates a semi-permeable membrane that at once links and separates the spaces of production and consumption (while not architectural in the same way, the delivery truck implies a similar possibility of linking production and consumption). The represented characters reinforce this division as the French pastry chef is aligned with the retail institution in an odd parallel between otherwise incongruous cultural signifiers: stereotyped Frenchness and Italianness being linked by a cultural imaginary of the culinary profession. Here, represented identity becomes one of many elements of symbolic bricolage that frame dramatic role-play. This characterizes identity as role-play, simultaneously leveraging Frenchness to imply a certain thematic performance and, in so doing, identifying Frenchness according to such performances. Similarly, the spatial bifurcation of production and consumption is gendered, as the set maintains a cultural masculinization of production and feminization of consumption indicated by the male chef and female customer. As socializing toys, such set designs facilitate role-play by materializing social norms and stereotypes that associate certain performed roles with certain types of people.

Despite the multilayered representational politics in this set, however, a metalevel reading of this set suggests relatively little about the playing child, who is not assumed to be male or female, French or Italian. This is where such thematic sets—even the ones marketed primarily toward boys—differ from LEGO Friends,[6] which genders not only the representation of identities within the thematic role-play but also the playing child, completely transforming how the symbolic bricolage of the medium functions.

The most notable instance of this is the LEGO Friends color scheme. Although LEGO rather weakly attempted to refute the "pink bricks" critique by arguing that "Pink bricks and elements have been included in LEGO sets for decades. The new colors introduced to create the LEGO Friends collection are two blues, two purples and two greens, based on

[6] Halfway between these poles, the Paradisa product line (1992–1997) was primarily marketed toward girls but plausibly fit into the world of the other thematic sets.

global research that indicated a wish for a bolder, more vibrant color palette to create the most interesting models" (Nipper 2012, n.p.), this statement admits what is immediately obvious to anyone: that LEGO Friends employs a distinctive color scheme expressly to market to girls. Consequently, this design choice adds a second, more metalevel layer of socialization to a toy that now not only invites players to play out the represented theme but also to play out being a girl consumer. Whereas the targeted marketing of playsets to boys is contained exclusively in boy-centric thematic presentations and paratextual advertising, the feminization of LEGO Friends runs much deeper—its unifying palette marks the entire line, regardless of theme, as "for girls."

Thus, whereas Pizza To Go culturally appropriates the colors of the Italian flag to enhance the thematic presentation of its pizzeria, the Heartlake City Pizzeria set (Set #41311) is an eye-searing mishmash of a red, white, and green Italian flag interspersed with the Friends pink, purple, and blue. And Stephanie's Pizzeria (Fig. 3.4) does not even bother with Italian colors but is rendered exclusively in the Friends hues, signifying not pizza but rather Stephanie's hot-pink aesthetic. Consequently, whether the theme is food service or animal care or even STEM learning (see Olivia's Invention Workshop, Set #3933), there are no LEGO Friends sets whose theme is not also "for girls." LEGO Friends, that is, facilitates a doubled role-play, scripting how roles are expressed within the staged drama and identifying a gendered role for the player herself.

Unsurprisingly, this signaling is not fully dissociated from the thematic presentations of the sets, which also skew toward feminized play themes. When contrasting surveys of non-Friends and Friends sets, for instance, Reich et al. (2018) discover eight highly gendered trends in LEGO set design and marketing:

"Themes 1 and 2: Men Have Professions, Women Have Hobbies" (p. 289)

"Theme 3: Being Male Involves Danger, Saving People, and Sense of Urgency" (p. 292)

"Themes 4 and 5: Females Want to Socialize; Males Enjoy Solitary Activities" (p. 293)

"Themes 6 and 7: Men Are Experts; Women Are Novices" (p. 293)

Fig. 3.4 Promotional image for the 2015 Stephanie's Pizzeria set (Set #41092). The lone red stripe in the table suggests the iconic color scheme of an Italian Pizzeria while the rest of the coloring (the purple-and-white stripes above the counter window and the pink base and seats) signify the assumed femininity of the LEGO Friends line. Although the different pieces of this fragmented stage are all presented as mid-action (pizzas in the oven and on the table, money on the counter), Stephanie herself is posed outside the action, facing the audience. This reinforces the notion that she has no predetermined role within this playspace but may instead participate in any of the implied roles

"Theme 8: For Women, Beauty Is Important" (p. 294)

The eighth theme concerning beauty inflects not only the thematic presentation of beautification but the aesthetics of the entire product line, as mentioned above. Themes 1, 2, 3, 6, and 7, moreover, are linked by a larger gendered ideology concerning the relationship between work and leisure that is not only present in various themes, but also speaks to how LEGO genders play itself.

Work and leisure are both at play in Stephanie's Pizzeria (Fig. 3.4), a small, architecturally deconstructed pizzeria set that manages to simultaneously distill and expand on the basic design of Pizza To Go. The largest element in this set is the counter window that recreates the toy theater frame without any pretense of any further architecture. This set maintains the core play features of Pizza To Go (counter, oven, delivery vehicle, table), losing only a sink but adding seating, cash, a telephone, salad, and drinks. While the deconstructed architecture is more characteristic of this set being on the more inexpensive side of the product line, the overall increase in playable domestic accessories[7] is characteristic of the Friends line. Like Pizza To Go, this set facilitates role-play within the pizzeria theme by staging both production and consumption. Unlike Pizza To Go, the character in this set does not reinforce such role-play. Stephanie wears the white and pink clothes that are her trademark, indicating that her identity is not linked with the signifiers of this professional environment, implying what Reich et al. (2018) noted: that in LEGO Friends, women are hobbyists and novices rather than professionals.

Despite the phrasing of the title "Stephanie's Pizzeria" implying ownership, nothing in this set identifies Stephanie, clad in her trademark non-work outfit, as a proprietor. Instead, Stephanie is presented as floating freely amidst a deconstructed workplace environment that implies a whole host of possible roles—chef, salesperson, delivery girl, consumer (Stephanie even patronizes another pizza establishment in the television spot "Pizza Night at Stephanie's"). This is common for the Friends characters, seemingly middle-class girls who move freely between various work environments and decidedly upper-class leisure activities,[8] strongly suggesting that their job is simply another leisure activity. Although this smacks of upper-class dilettantism and, to a lesser extent, Barbie's professional cosplay,[9] I argue that the primary significance here is the

[7] The cultural feminization of accessories of fashion or domesticity (toys for boys often feature weapons or tools) recurs multiple times in LEGO history as LEGO has released several lines of buildable jewelry and other accessories directly marketed towards girls, such as Scala (1979–1980) and Clikits (2003–2006).

[8] This is reminiscent of the Paradisa line (1992–1997), which also feminized leisure activities.

[9] Traditionally, Barbie does have a job—she is a teen fashion model—but she does not hold most of the jobs she dresses as.

intermingling of work as leisure that subjects the Heartlake City economy to a logic of *play*.

In this world of play, various establishments serve as backdrops for the girls to engage in a kind of participatory role-play that aligns them with the implied girl consumers vicariously mediating their own role-play through the LEGO Friends toys. This challenges a traditional function of socializing toys: preparing children for real-world professional interactions. Or, rather, it redirects this socializing function. If girl players are not being socialized as aspiring professionals, they are being socialized as consumers, reinforcing the expectation that as homemakers women need to be fluent in a domestic economy that involves interacting with many professions that the consumer does not actually perform. At the same time, emphasizing such performances as play may incline the socializing messages more toward the relational than the representational, suggesting that girls are playacting at playacting itself, thereby developing social skills in performing culturally-constructed gender roles.

This sociality undermines the representation of work as work, construing work as a playful performance rather than a means toward economic productivity. This would not necessarily be problematic—I am all for sociality undermining productivity as a radical alternative to work-obsessed capitalism,—if it were not so clearly gendered. When targeted directly toward girls, any potential anti-capitalist message quickly becomes reconstrued as a variant on "Men Have Professions, Women Have Hobbies" (Reich et al. 2018, p. 289). Consider, for instance, the often-touted exception that proves the rule—Olivia's Invention Workshop (Set #3933), a rare STEM-themed LEGO Friends set. While genuinely celebrating Olivia's technical proficiency, this small set—once again in the feminized Friends color scheme—still suggests home hobbyism rather than any professional context.

When women do work in LEGO Friends, their work is vastly skewed toward forms of *affective labor* (Hardt 1999) that have been traditionally feminized and, consequently, devalued in Western society. Even Olivia's Invention Workshop domesticates technical proficiency by foregrounding Olivia playing with and caring for (with the provided oil can) a small, cute robot. Indeed, the history of identifying mothering and housekeeping as archetypal forms of so-called 'women's work' is reflected in the two of the most popular forms of affective labor in LEGO Friends: animal care (see Merskin 2019) and food service. For instance, the product blurb for Heartlake Vet (Set #3188) reads "Today, Sophie and her assistant Mia need to examine Bella the horse, Scarlett the dog and Oscar the hedgehog to diagnose and cure these cute little critters." As this description implies, although this set promises "lifelike vet accessories" that make it "both educational and fun," the design aesthetic has a cutesy rather than clinical tone. While Sophie, a doctor who is never addressed as "Dr.," is a working professional and not a hobbyist, nothing in this description or set reinforces an aura of professionalism. As an aspiring professional, "her assistant Mia" is not exactly playing doctor, but she is genuinely learning a type of professionalism that is thoroughly dependent on a feminized ideology of affective labor.

The set designs of both the early thematic playsets and the later 'for girls' LEGO Friends sets stage gender by architecturally scripting the types of identity play that these physical stages facilitate. While the embedded scripts in these playsets cannot guarantee any particular form of actual play, they establish an ideological context that thoroughly inflects the mediated meaning-making system within which dramatic play takes place. Unfortunately, this ideological construction often reinforces problematic social norms. The early thematic playscripts rely on iconic significations that favor stereotypical identities over nuanced identities. And even more problematically, the LEGO Friends sets feminize even the act of dramatic play itself.

Character Design: Signifying Sex, Performing Gender

Having set the stage for dramatic play, these ideologically-laden play sets invite players to perform actual toy dramas. This immediately involves a second, more direct form of mediation—the ideological construction of identity built into the bodies of the toy characters. The role of toy characters varies quite dramatically across LEGO history from the archetypal social roles embodied by the minifigures in Pizza To Go to the named LEGO Friends characters with personality quirks and interests who are presented more as friends than roles to be played. While gender is clearly present in both modes of characterization, these modes are themselves gendered with the sociality of the Friends being clearly feminized. Thus, just as "It is gender … that the modern doll primarily *does*" (Driscoll 2015, p. 187), the comparative design of LEGO's minifigures and mini-dolls itself embodies a gendered dynamic in which the masculinized minifigures are more closely aligned with construction play and the feminized mini-dolls are more closely aligned with dramatic play.

Looking to figure design is crucial for deconstructing the ideological formation of LEGO, which has always paralleled innovations in figure design with innovations in toy-mediated storytelling. Indeed, it was the introduction of the minifigure in 1978 that catalyzed the development of LEGO around *play themes*, including the classic Space and Castle themes. Shortly thereafter, LEGO produced Fabuland (1979–1989), a whimsical world of anthropomorphic animal figures that introduced named characters who starred in storybooks and a stop-motion television series. Since then, LEGO has conducted many experiments that introduce new figures to anchor narrative product lines, including the relatively short-lived Jack Stone (2001–2003) and Galidor (2002–2003) lines. Despite these missteps, LEGO clearly understands that dramatic play is heavily dependent on character toys to provide affective loci for players.

LEGO Friends is only possible because redesigning character toys can completely transform the feel of a narrative world. Aside from the buildable figures of Bionicle (2001–2010, 2015–2017), which revolutionize narrative play by constituting a new construction system unto itself, LEGO Friends has arguably become the only successful, lasting pairing between a new figure design and a play theme. Whereas the material designs of the unnamed minifigure characters in Pizza To Go were clear generic extensions of the same symbolic bricolage that characterized the overall set design, the mini-doll characters in LEGO Friends have a much more nuanced expression of embodied identity. After all, embodied toys simultaneously provide affordances for dramatic play and evoke complex body politics that are often culturally gendered. Whereas the former thematic style simply equates embodied identity and role, the latter treats embodied identity as an expression of internal identity. In a complex identity play, LEGO Friends mini-doll bodies (Fig. 3.5) express both biological sex and performed gender as they materially embody their own dramatic performance.

This identity play can be elucidated by the work of Judith Butler (1990/2008), who confronts discourses of biological essence to argue that identity unfolds as a kind of embodied performance. To clarify the binary opposition between male and female, Butler distinguishes two quite different forms of embodiment: *sex* and *gender*. Sex is a purely biological distinction, pertaining primarily to the two sex's differing

Fig. 3.5 A side-by-side comparison of a LEGO minifigure and a LEGO Friends mini-doll presented as part of a press release packet accompanying the release of the LEGO Friends line. The differences in proportion are measured according to a specifically LEGO background grid that shows increments in the height of LEGO bricks (and sub-increments for the height of LEGO plates, which are a third of a brick). The mini-doll is therefore exactly one brick taller, despite being significantly slimmer in the body. Also note that the mini-doll has an additional point of connection for hair accessories, a three-dimensional skirt that makes it impossible to move the legs independently, and hands that are in fixed positions

yet complementary reproductive roles. Gender is, by contrast, something people *do* rather than something people *are*. In other words, 'sex' names the distinction between (biologically) male and female whereas 'gender' names the distinction between (culturally) masculine and feminine. Biology may inform this performance to some extent, but—at least according to Butler—much less than one might expect.

Obviously, the question of biology does not directly apply to toys whose bodies consist of molded plastic. Yet, as material objects that anchor performances, toys both *embody* and *perform* identity in ways that parallel certain aspects of Butler's distinction between sex and gender. This section departs significantly from the substance of Butler's analysis to instead pursue a parallel comparison between the embodied material design of toys (their putative 'sex') and their role within dramatic play (their putative 'gender').

Biological sex is an odd notion for toy bodies, which almost universally lack genitalia and, more importantly, universally lack any biological characteristics whatsoever. Strictly speaking, toys lack sex altogether—they are inanimate objects, the neuter 'it.' And yet, character toys are typically unambiguously labeled as 'he' or 'she' rather than 'it.' Although it might seem obvious that character toys need some way of representing sexed characters without being sexed themselves—after all, photographs of biological humans also have no sexual characteristics themselves—what is important is *how* toys use their media-specific characteristics to represent sex. In particular, I suggest that toys almost invariably perform gender in ways designed to stand in for biological sex. Toys, in other words, play with sex and gender in ways that significantly depart from the gendered performances of biologically sexed bodies that Butler analyzes.

Thus, the most common paradigm of representing sex in contemporary toys is a *hypersexualization* of form through the exaggeration of certain idealized body proportions—the exaggerated slimness, hourglass figure, and breasts for female dolls like Barbie and the exaggerated muscularity of male action figures like G.I. Joe. Such hypersexualization typically references cultivated characteristics that contribute to 'sexiness'—the kinds of figures that the gendered beauty industry encourages people to emulate and adore. Ironically, these toys hypersexualize their characters without referencing any biological sexual characteristics such as reproductive organs, which are smoothed over. As a consequence, these toy designs embody less biological sex itself and more culturally-constructed ideals of sex or sexiness.

Departing from this tradition, the LEGO minifigure's representation of sex is highly unusual in being completely physically unmarked. While sexual difference is sometimes alluded to, it is given no tangible existence—to the touch, LEGO male and female minifigures are indistinguishable and are both more reminiscent of the blocky medium itself than any human figure. In other words, the material form that signifies the 'biological' make-up of the LEGO minifigure is distinctively architectural, as if LEGO minifigure characters were the personified spirit of the construction medium itself.

Instead, the most notable way sex is marked on LEGO figures is through printed lines that imply curves. These lines evoke bodily characteristics without being materially embodied and even in a sense belying the material reality of the figures themselves. For instance, more recent female minifigures have printed curves that imply breasts or an hourglass figure, a topography that is imaginatively projected beyond or within a smooth, geometric solid. Implying respectively an absent presence and a present absence,[10] these printed bodily features signify but do not materialize sexed body characteristics, thoroughly complicating what it means for a toy minifigure body to be sexed. Notably, this sexed projection onto the architectural LEGO body is exclusive to femaleness.

Maleness, by contrast, is so strongly implied as the norm in the *androcentric* (Bem 1993) world of the LEGO construction toy that it is often not signified at all (the most common exception being facial hair).[11] On this point, the delivery *boy* in the Pizza To Go set is particularly telling. This character is, in fact, completely unmarked with respect to both sex and gender, featuring a completely unisex outfit and the ubiquitous LEGO smiley-face.[12] The only maleness this character has is the weight of conventional expectation, which is strong enough that it is likely to pass unquestioned.

Whereas Barbie bears her sex in her body, her race on her skin, and her constantly shifting occupations in her modular wardrobe, LEGO minifigures bear all three simultaneously at the boundary that brings together

[10] I find this latter absence particularly disturbing as the printed cutouts that frame the hourglass figure imply that the female body should take up less space than the figure actually takes. While Barbie is often criticized for having an embodied ideal that is physically unrealistic, the LEGO minifigure actualizes the physical differential between the body and the ideal. While I would not suggest that such representation leads to anorexia, it certainly symbolically replicates a form of anorectic mindset.

[11] This makes the scene in *The LEGO Movie 2* when Emmet transforms himself into Rex Dangervest (see Chapter 6) particularly significant, because he supplements his wardrobe and name changes by drawing stubble on his face with a permanent marker. Whereas for a normal human, stubble represents a period of transition between clean-shaven and full-bearded, the non-biological LEGO minifigure reimagines this biological state as a fixed signifier of rugged masculinity.

[12] Initially, the yellow smiley-face was designed to be a blank slate upon which children could imaginatively author their own significances (much like the missing mouth of Hello Kitty, which was intended provide space of potentiality where children could create their own expressions). LEGO tells the story this way: "When the minifigure first appeared, it was decided that its face should have only one colour: yellow. And that its facial features should be happy and neutral. The figure would have no sex, race or role – these would be determined by the child's imagination and play. It was not until the launch of LEGO Pirates in the 1980s that figures who could be evil or good, happy or grumpy were developed and launched" (2014, p. 14).

body, skin,[13] and wardrobe. As LEGO weaves these three facets of identity together in a single surface presentation, the most common ways of differentiating male and female minifigures are universally cosmetic, located exclusively in the ink printed onto the bodies. And, beyond this visual implication of a materially non-existent bodily difference, the most common of these is literally cosmetic: the lipstick (and, secondarily, eyeliner) that identify all feminized minifigures as female. For such minifigures, femaleness becomes something that cannot inhere in bodies but can only be layered onto bodies via a gendered performance.

While signifying maleness as default and femaleness as a cosmetic performance is already problematic, it becomes more so due to the ways minifigures come to represent the medium. Because identity is representationally layered only onto the surface of minifigure bodies, their blocky aesthetic is more indicative of the brick-based medium itself than any sexed or gendered bodily characteristic (including the blockiness of masculinized muscularity). To put it another way, the genetic ancestry of the minifigure is not the intersection of male and female antecedents but rather the coupling of construction toy and character toy—the LEGO minifigure is at once brick and doll, the anthropomorphized form of the medium itself. Pulling these pieces together, the ideological construction of LEGO minifigure design assumes maleness as the default gender identity, identifies LEGO bodies as the anthropomorphized projections of the medium itself, and thereby genders construction play itself as masculine.

Conversely, the mini-dolls (Fig. 3.5) launched in 2012 as the new feminized characters in LEGO Friends gender dramatic play as feminine by less clearly reflecting the medium and more clearly reflecting traditional dolls. Unlike the bodily uniform minifigures, these slim, curvy mini-dolls exhibit three distinctive body types corresponding to different biological sexes (and ages): young female, adult female, and male. These bodies are more sexed than sexualized, as their bodily differences are extremely slight compared with gendered toys like Barbie, at least in part because

[13] Although in the interests of time I have chosen to explore the identity politics of gender rather than race in this project, it is an undoubtedly fruitful comparison and one that could use significantly more scholarship. Neither case is simply a matter of representation (although representation certainly matters). Just as the representation of gender accompanies a gendering of dramatic play, associating the supposedly 'neutral' yellow minifigures with whiteness also reflects a centering of Western visions of play culture. In other words, the ideal of play built into the LEGO medium is not just about who can participate—after all, it is in LEGO's financial interests to reach a diverse market—but more subtly about what cultures of play are privileged in the design. See also Johnson (2014b) and Cook (2017).

the world of LEGO Friends is decidedly child-centric.[14] The sharpest differentiation, therefore, is not between female and male mini-dolls, but between masculinized minifigures and feminized mini-dolls. Whereas minifigures resemble the blocky medium itself, mini-dolls reflect familiar doll aesthetics, merging the slimness of the fashion doll with the cuteness of the baby doll, with soft curves and large eyes designed to elicit feelings of emotional attachment.

Again evoking the stereotype that "Females Want to Socialize," these characters are designed less to interface with the medium and more to function as traditional character toys, evoking the affects and affectations of dramatic role-play. In other words, minifigures emphasize action and material connection whereas mini-dolls emphasize aesthetics and social connection. More specifically, somewhat reflecting the action figure tradition, LEGO mini*figures* feature more points of articulation than most dolls despite being slightly smaller. The tableau on the Pizza To Go box, for example, depends on the high degree of pose-ability offered by these figures. Note the degree of narrative contained in the relative positioning of the chef and patron, whose eyes interlock as the patron impatiently leans in and waves in the harried, food-laden chef. In contrast, Mini-dolls have three fewer points of articulation than minifigures, making them significantly less pose-able, but have an additional point of connection on the hairpieces that allow for accessories to be added to heighten the aesthetic impact.

Interestingly, whereas minifigures are differentiated exclusively via surface printings (on which body type, skin color, and fashion choice are all inscribed), mini-dolls use bodily differences for more than just differentiating body types. Mini-doll design extends the representation of fashion beyond printing, producing different leg molds to represent different clothing styles (Stephanie's shorts, for example, are part of her body). The result of these design changes is that mini-dolls locate gendered distinctions *in* the body (as sexed characteristics) rather than only *on* the body (as the minifigure does). Here, like the minifigures, body and fashion have ontologically equivalent status in the mini-doll body. Yet, unlike minifigures, these distinctions are not only on the surface but are

[14]While toys like Barbie and G.I. Joe are undeniably *for* children, the characters are not themselves children. Instead, these functions play out fantasies of occupying more grown-up roles (Joe is obviously an adult and Barbie is a teenager, but one who is rendered more adult-like by her professional career as a model who primarily dresses as other adult professionals). The Friends, by contrast, are somewhat indeterminately aged from late childhood to early teen.

also in the form of the body itself, encoding identity at a deeper, more tangible level. Thus, the mini-doll performs gender as an essential as well as a cultural characteristic, presenting sex and gender as alike constitutive of identity in a way that is more aligned with the understanding of gender that Butler critiques (a biologically essentialist argument that sex should dictate gender) than the one she proposes (a non-essentialist argument in which culture primarily dictates how people perceive both sex and gender).

Consequently, just as the set design in LEGO Friends represented not only particular themes but also a world of girls' play, the character design of LEGO Friends mini-dolls encode more fixed identities into their bodies and are less flexibly interchanged and intermixed. With their distinctive bodies—especially following a 2018 redesign to make the Friends more ethnically diverse—personal names, and archetypal personalities, the Friends are meant to function much more as characters in staged dramas than the LEGO minifigures which are more clearly designed as sets of generic signifiers meant to be remixed (albeit more in the sense of LEGO bricolage than *trans** identity[15]). Consequently,

[15] This remixing allows for intriguing bodily play, as Jack Halberstam discusses in *Trans**, building on the fact that "Lego architectures are in a constant state of emergence and collapse" (2018, p. 131) to call for using "the moveable parts and freestyle building processes associated with Lego to think seriously about new bodily architectures, component parts of embodiment, and structures of becoming" (pp. 129–130). Yet, it is worth noting that Halberstam's exploration is woven from two threads—the process of LEGO architecture and the narrative of *The LEGO Movie*—that do not include the reality of minifigure bodies. Minifigure bodies are not trans* bodies, that is, bodies that "represent the art of becoming, the necessity of imagining, and the fleshly insistence of transitivity" (p. 24). While minifigure bodies do represent "the necessity of imagining," becoming and transitivity are not reflected in how the highly-symbolic LEGO bodies present identity through formulaic and permanently inscribed signifiers. Unlike a human body that may transform in many ways, whether through the unavoidable transformations that define a biological lifecycle or more intentional medical interventions, the LEGO minifigure has no way of maintaining an identity that spans this transition. That is, whereas a human body may literally *transition* from one state to another and gain meaning not only from the new state but from the process itself, a LEGO body cannot. As one part completely replaces another, one bodily configuration completely erases or replaces another. To put it another way, LEGO bodies jump from one assemblage to another without occupying or bearing traces of any transitional state. This transition simply reifies the binary, undermining the performative aspects of minifigure design. For Butler, although embodied sex does not causally determine performed gender, embodiment—with its int(e)ractable profundity—precedes the performances it enables in ways that call into question any one-to-one relationship between bodies and identities. Minifigure bodies, by contrast, take

the masculinized minifigures are construction toys clearly designed to reflect the medium whereas the feminized mini-dolls are character toys designed to star in dramatic play that is staged upon constructed playsets, but which do not easily interface[16] with them.

Thus, while no biological essence determines the ideological and cultural construct that is gender performativity, the material design of these character toys manage to embody these ideological and cultural constructs themselves. Consequently, toy bodies determine toy identities—and thereby script performed identity play—much more thoroughly than human bodies do. These already-performative bodies call out for performance. And, in so doing, they ask players to perform their gendered identity construction through their dramatic play.

Paratexts: Performing Sociality

Together, the theatrical stages of the scripted playsets and the materialized identities of the toy characters provide a strong ideological framework for dramatic play, providing an abundance of material for performing and scripting actual play. Taking this still further, LEGO has increasingly supplemented these embedded scripts with circumscribing paratexts that double down on their gendered ideological constructions. Just as toy theaters often explicitly adapted famous stage plays to elicit familiar scripts for dramatic play, LEGO creates promotional media content that dramatizes playful narratives in ways that further suggest how children should play with their already-scripted toys.

In this vein, the detailed storyworld of Heartlake City functions simultaneously as a setting and a script for the feminized dramatic play of LEGO Friends. Much more developed than a typical LEGO play theme, Heartlake is an expansive world articulated through LEGO playsets, informational paratexts like maps and character bios, and narrative paratexts that relate domestic dramas in multiple media. While the significations

the social signifiers of gendered performances and write them on the bodies, concretizing performed gender to the point where it functions with the absoluteness of biological sex.

[16]LEGO minifigures have two ways of connecting with standard LEGO bricks that mini-dolls lack: connections on the back of the legs, and connections inside the torso (both have connections on the bottom of the feet and atop the head). The main consequence of this is that mini-dolls rest atop rather than connect to their environment when sitting, heightening the impression that they are dolls in a dollhouse rather than part of the medium itself.

built into the architecture of the city itself are already highly suggestive, these paratexts provide explicit narratives that can function as direct models for dramatic play. In particular, these narratives directly model the feminized ethos of performing sociality already present in the toys' implicit playscripts.

Sociality is established first and foremost by situating the player within the social space of Heartlake City. Unlike the cutout characters that typically star in toy theater dramas, two-dimensional façades that function like masks for theatrical performances, LEGO Friends strongly implies that the human players should think of themselves as *friends* of the Friends. Although it is not uncommon for toys or dolls to imply companionship with their players, many toys structure relationships in other ways. For instance, some toys invite playing *as* a particular character (toys as avatars), some invite playing out stories that do not include the player at all (toys as theatrical props; see this chapter), some (toy soldiers, baby dolls, the pedestrians in the LEGO Town Plan; see Chapter 2) invite other non-friend relationships, and so on. Whereas even some stereotypical feminized dolls show considerable variation within these possibilities,[17] Friends is very consistent about positioning its players as members of the sociality they are invited to dramatize.

For instance, the back of the instruction manual for the flagship set from the initial 2012 Friends line, Olivia's House (Fig. 3.6), is a clear invitation to the girl consumer to join the collective of Friends as a participant rather than a puppeteer. Whereas many other LEGO sets break the fourth wall to make their playsets more accessible to theatrical puppeteering, this advertisement breaks the fourth wall to establish sociality by making the Friends directly smile at and make eye contact with the consumer. Olivia holds out a dollhouse-style set representing her own house to the viewer, simultaneously inviting the viewer to share in playing house and to metonymically imagine being invited to her house. By mixing the diegetic and extradiegetic worlds, showing the Friends playing and displaying their

[17] Although many dolls (including Friends and American Girl dolls) are often positioned as companions to young girls, the most famous doll—Barbie—is more variable. As a fashion doll, Barbie functions as a model for the player's fashion play. And as a doll often accompanied by various forms of professional cosplay, Barbie can also become an avatar for aspirational play.

Fig. 3.6 Promotional image from the LEGO Friends Olivia's House set (Set #3315). All five of the titular Friends make smiling eye contact with the viewer while Olivia holds out the playset of her house to the audience (the three-dimensionality of this is heightened by the slight white haze that makes the product seem to pop out from the arrayed Friends). The invitation to play with this dollhouse set is therefore framed as an invitation to participate in the world of the Friends as a sixth friend (note that the exterior of the house is shown as if the audience are guests who are invited through the front door into the domestic space inhabited by the Friends, who are layered in a way that they seem to radiate outward from the implied interior space)

own toys,[18] this image suggests that playing house in LEGO is a fundamentally social activity—both thematically (insofar as its dramatic play involves staging social situations with the Friends and family that populate Heartlake City) and performatively (insofar as the dollhouse is positioned as the site of social play with friends and family). While this sociality can be quite sweet, it nonetheless reinforces the confining stereotype that "Females Want to Socialize" (Reich et al. 2018, p. 293). Sociality, in other words, is treated here as a feminine rather than universal value, which is especially unfortunate given that sociality *is* actually treated as a universal LEGO value elsewhere (see Chapter 6).

Looking more broadly, varied invitations toward sociality are refracted across the many spinoff themes that make up the Friends sub-medium, including DC Super Hero Girls (2017), Elves (2015–2018), and Disney Princesses (2014–present). These lines each refract the distinctive aesthetic and ethos of Friends in their own way—the DC and Disney lines merge the Friends aesthetic with that of their adapted source texts, while Elves offers a more ethereal, fantasy-like aesthetic. Yet the consistent ideological threads that run between these superficially distinct lines suggest that 'for girls' LEGO truly does have a prevailing ideology. In particular, these lines all feminize dramatic play in ways that treat performing sociality as the primary means and ends of LEGO play for girls.

These lines firstly feminize dramatic play by emphasizing plotlines predicated on sociality. Certainly, one might expect the main LEGO Friends line to elicit domestic dramas that emphasize *friendship*. Yet, this emphasis is not at all exclusive to the domestic storyworld of Heartlake City—both the comic book adventure DC Super Hero Girls and fantasy LEGO Elves themes domesticate these adventure genres to instead emphasize sociality. Set as a whimsical high school version of the canonical DC storyworld, DC Super Hero Girls,[19] appropriates the fan fiction genre of the *alternate universe* to reimagine the epic DC comics universe

[18] Similarly, several of the television advertisements cut between live-action shots of hands playing with the Friends toys and animates scenes where the Friends come to life and even talk with the viewer.

[19] This line is a LEGO adaptation of an existing 'for girls' line within the DC comics universe. LEGO also has an independent DC Comics Super Heroes line. According to the Brickset database, there are 10 sets in the main DC Comics line for every one set in the DC Super Heroes Girls line.

as a more relatable youth drama. In many ways, this series feels like a comedic mashup of a superhero comic and LEGO Friends.

Whereas DC Super Hero Girls reimagines adventure heroines in everyday social settings, LEGO Elves transports an otherwise normal young girl named Emily into a fantasy world. Interestingly, while Elves reproduces the standard adventure narrative of many boy-centric LEGO lines that feature a collection-oriented quest in which the protagonists must collect a full set of something-or-others (a storytelling strategy that aligns neatly with the marketing of individual sets into larger product lines), much of the plot hinges on Emily—whose elemental power is love—befriending different elemental elves. In contrast to LEGO Ninjago (2011–present), where the elemental ninja are already a team and the play focuses on their adventures and battles, Elves pays more attention to the encounters that establish sociality with the various inhabitants of the fantasy world. Consequently, like the DC Super Hero Girls, this line suggests that sociality is a central weapon of the feminized heroine and, by extension, a central scripted value for the feminized player.

As this storytelling emphasizes feminized dramatic play, these themes also often deny *construction* as a central theme. The proprietary[20] LEGO storyworlds of Friends and Elves are especially prone to this, typically animating their stories as if they were not related to LEGO in any way. While the style can vary, these animations are so visually rooted in familiar two- and three-dimensional animation traditions that an outside observer might watch an entire video without realizing that these characters have LEGO counterparts. This denial that construction is central to the LEGO brand is a sharp contrast to other LEGO animations, including those discussed in Chapters 5 and 6, which often visually emphasize the theme and aesthetic of LEGO construction, even when they also dramatize sociality.

Such videos rely on the medium of animation to create character dramas completely unhinged from the LEGO medium, privileging dramatic play over construction play by visually denying any connection to a construction toy. More interestingly, the paratextual videos that do embrace toy-centric storytelling still manage to privilege dramatic over construction play. This is most clearly shown in the Disney Princess line,

[20]While the DC Super Hero Girls and Disney Princess lines are more distinctly LEGO-like, that may be an artifact of the necessity of branding these lines as LEGO, since both media franchises already feature non-LEGO animations.

which animates its characters primarily as toys. Yet, unlike most *toys-to-life* narratives, these videos animate toys in ways that strongly imply young girls' dramatic play.

One series features mini-doll characters starring in short recaps of their source films. Although the mini-doll heroines are animated as three-dimensional characters, the videos are staged like toy or puppet theaters that mix brick-built structures with cardboard cutouts that have a decidedly DIY aesthetic. Aside from the heroines, most other characters are also hand-drawn cardboard cutouts on sticks that give the impression of being puppeted from outside the frame. Reinforcing this theatrical staging, a young girl narrator retells the story in a highly condensed and comic way. Consequently, these videos flaunt the toy status of their animated characters as a way of advertising their potential for dramatic play. Rather than offering stories unto themselves, they model and thereby advertise the practice of using licensed character toys to retell canonical stories. As these films primarily feature the relational drama of coming-of-age romance plots, these minisodes function as paratextual playscripts that direct an already feminized dramatic play toward further feminized romantic plotlines.

While these recap minisodes advertise the playful retelling of canonical storylines, other Disney Princess videos star the Disney Princesses having adventures within a child's bedroom. These toys-to-life videos remove the toy theater frame and instead animate the characters as if they had come to life. Yet, contradicting this visual animation, these videos maintain the singular voice of a young girl narrator, suggesting that these characters are expressions of a singular imagination. Whereas the recap minisodes emphasize the romantic plotlines of the source texts, these toy stories emphasize female bonding.[21] As the Disney Princesses

[21] This is an important evolution in the Disney Princess franchise. Disney Princess is a very peculiar media franchise pieced together from similarly branded storyworlds that do not narratively overlap. As Peggy Orenstein notes, "princesses avoid female bonding" (2011, p. 23) to such an extent that when they appear together in Disney advertising, they "won't even *look* at one another" (p. 23). Yet, whereas the narratives that the Disney Princesses hail from do not provide an easy way to bring them together, their status as branded toys does, which is why LEGO is able to easily narrate this kind of female bonding or why *Wreck-It Ralph 2* (an animated meta-narrative about the lives videogame characters live behind the screen) is able to bring the Disney Princesses together. As time has passed, Disney has continued to explore the complex dynamic between these overlapping yet not-overlapping worlds in various ways that is well-deserving of further research.

abandon their canonical plotlines to instead participate in stereotypical female bonding plots that resemble those in LEGO Friends, they bring the Disney Princess line back into the primary ideological construction of the Friends sub-medium, a feminized toy for dramatizing sociality.

Consequently, whether these paratexts animate dramatic play by denying their toy status altogether, animate theatrical retellings of canonically feminized film narratives, or animate toys as ways of playing out imagined narratives of female bonding, these varied animation styles all point toward a single message: that girls' play is defined by dramatizing sociality. As all these products are visibly marked and marketed as 'for girls,' this reveals an ideological assumption that girls inherently prefer dramatic play over construction play. And while this may hold true in many individual cases, playscripts transform assumptions into expectations, actively constructing the gendered oppositions they assume. As these advertising messages are entirely predicated on selling the play they model, they necessarily weave scripted, socializing messages in and around these toys in ways that encourage girls to value dramatic play over construction play.

A House Divided

That a popular contemporary toy would be heavily gendered is, unfortunately, nothing unusual. What is unusual is that LEGO stands as a house divided, a single medium ideologically bifurcated by its gendered messages. Whereas many other toys have embraced gendered market segmentation by relentlessly pursuing a single demographic, LEGO advertises itself as a single universal system internally split to separately target both sides of a gendered divide. As a consequence, the construction of girlhood—and, by extension, womanhood—in these sets positions female identity as both an exception and supplement to male identity. And, in so doing, this gendering of LEGO play suggests an internal hierarchy within the medium itself, such that the supposedly neutral 'standard' or 'classic' LEGO is implicitly assumed to be for boys whereas the 'for girls' LEGO Friends line is explicitly assumed to be gendered.

This gendered ideological message contradicts the material reality of the toys themselves. Despite activist claims that construction was overly simplified in the Friends redesign, LEGO's somewhat defensive claim that consumers of Friends "enjoy the exact same building experience and developmental benefits" (Nipper 2012, n.p.) was partially confirmed by

Black et al., who note that "Combining the results from the three analyses offers some evidence that, in terms of the complexity of their products, LEGO is treating girls as more developmentally advanced than boys" (2016, p. 71). Consequently, the issue is not that LEGO Friends fails to offer construction play to girls, but rather that it offers this construction play within a value system that implicitly genders the forms of play it offers and, thereby, bifurcates the medium into an implicitly masculinized standard and an explicitly feminized sub-medium. At stake in gendered LEGO toys, that is, is a highly problematic philosophy of binarized gendered play. While it is certainly questionable to assume strong 'natural' differences between girls' and boys' play, this section explores another set of ethical problematics: the misrepresentation of construction and dramatic play as inherently oppositional and the bifurcation of the LEGO medium into gendered media designed to remain distinct despite their material compatibility. At stake in the gendered design and marketing of LEGO, that is, is an entire gendered theory of play.

That LEGO assumes a highly binarized vision of gender is unfortunately reflective of a common trend in gendered toy design. Reflecting on the semiotics of contemporary toys, Dan Fleming argues "That it is a male body at the centre, and not a female, now seems an inescapable fact" (1996, p. 52). He concludes, furthermore, that the female body in toys serves as an exploration of difference, one of the ways in which "toys appear to 'test' the nature of male identity by pulling it first in one direction, then in another" (p. 50). This is not to say that girls' toys are not actually 'for girls,' but rather that the desire to create a 'toy of one's own' necessitates that such toys dwell in contradistinction to boys' toys, which can be considered the norm because they do not have to express their identity in such clear opposition to girls' toys.

As one might expect, the binary that emerges from this gendering of LEGO toys is by no means 'separate but equal.' As Derek Johnson puts it,

> … these moves to expand gendered appeals within major media brands give girls "a room of one's own", yet simultaneously tell girls to stay in that gender-defined space so fully separate (and unequal) market emphasis on boys and men can be sustained. Historically speaking, cultural producers across numerous media have long reacted negatively to the idea of girls and women playing freely with and generating their own meanings from popular culture (in fandom, for example), regulating use of media and

> culture instead through gender conformity. LEGO's rebooted market suggests that girls' creativity should follow gender scripts, and simultaneously holds (or pretends) that they would not be interested in play patterns free of that governance. (2014a, p. 88)

In other words, LEGO reflects a line of patriarchal thought in which masculinization is interpreted as neutrality, and only feminization is marked as gendered.

Beyond merely signifying gender, this gendering is normative: more masculinized LEGO minifigure bodies are coded less as masculine and more as inherently LEGO whereas feminized LEGO mini-dolls are explicitly coded as feminine in ways that make them seem outside the LEGO medium. Thus, LEGO Friends is always presented as a special case, a protected space or state of exception that offers something different—i.e. feminized—from traditional LEGO. Friends is never presented as the next stage in the evolution of traditional LEGO nor are girl consumers presented as LEGO's core market.

Indeed, LEGO has consistently implied that LEGO Friends is not 'real' LEGO by, for instance, excluding the LEGO Friends from the LEGO Club Magazine (for a significant time, consumers identified as girls received a supplemental Friends insert literally glued inside the normal magazine). Thus, the gendering of toys that Fleming and Johnson note strongly reflects a theoretical argument by feminist philosopher Nancy Jay (1981), who argues that the binary logic of male and female is represented as the distinction between 1 and 0, between male and not-male, implying that female identity is constructed in contradistinction to masculinity rather than in itself. And yet, as Jay argues, because maleness is the assumed default, only deviations from the norm are marked as gendered, as how LEGO Friends cannot escape being continually and overtly feminized whereas traditional LEGO is more subtly and implicitly masculinized.

Consequently, Friends is not just another LEGO play theme like City or Ninjago. Friends is essentially a sub-medium like Technic or Bionicle; a whole new feminized variant on the LEGO meaning-making system. While many of its pink bricks are just classic LEGO with a new coat of paint, Friends offers many new elements—most notably its completely redesigned mini-dolls, animals, and accessories—that transform the construction toy by fusing it with doll play. While all LEGO has a core element of dramatic play, this distinct-yet-compatible sub-medium offers new material possibilities for explicitly feminized performances.

The upshot is that although construction play and dramatic play are thoroughly intertwined, the messages that surround LEGO play still drive an ideological wedge between these forms of play that presents some core elements of LEGO play as more intrinsically LEGO than others. And it does this in a way that directly contradicts the implicit messages of LEGO design, in which the prevalence of thematic playsets over abstract construction sets indicates that dramatic play is a central focus for the medium. Although in any individual performance of play these moments can and often do complement each other, the rhetoric that genders construction play as for boys and dramatic play as for girls imposes an artificial logic of separation on LEGO's branded self-concept.

As cultural critic Sarkeesian notes of the advertising for LEGO Friends, "the play time is supposed to happen after the building is complete, unlike in the other commercials where boys are encouraged to actively build and construct as part of their LEGO experience" (2012, n.p.). Similarly, Derek Johnson (2014a) notes how whereas LEGO designer videos for boys' lines involve discussions of actual design, female Friends designers are often shown playing with their designs. In these moments, LEGO not only scripts its dramatic play according to traditional gender roles, it genders itself as a play medium. Although dramatic play in LEGO is still a form of creative bricolage in which players author and stage their dramatic play, the material design of LEGO sets the stage for performing gender according to fairly traditional roles. As typically happens with LEGO, a tension emerges between the openness of the medium and the closedness of its messages, which coexist uneasily in the gendered performances that characterize LEGO's dramatic play. This contradicts bricolage as the underlying ideology of the LEGO medium by suggesting that although the system is designed to allow for disparate elements to be brought into creative combinations, players ought not to do so in gendered play.

Given that gender is inherently performative, there is special danger in feminizing performance itself, as toy scholar Stephen Kline observes in contemporary toy culture: "Girls' play, it seems, is more like a spectacle or performance than an act of imagination" (1993, p. 253). Whereas boys' play is typically described as self-actualizing—construction play facilitating cognitive development and dramatic play encouraging imagination, aspiration, and agency—staging girls' play as performance suggests social normalization, contributing to the politics of surveillance that characterized the aforementioned doll culture and complicating how the girl in the LEGO advertisement performs her self-actualization for her mother. In other words, girls' toys can be structurally objectifying even when they are not explicitly sexualizing, simply by being dramatically oriented toward display. In the most extreme interpretation, this implies that what it means to be a girl is to perform some non-essential identity, a fabrication that truly functions as the binarized 0 with respect to the essential identity of the masculinized 1.

Even in less extreme interpretations, one might question the wisdom of training girls to perform their play in these ways in a culture where "the girls I talk to respond to questions about how their bodies feel—questions about sexuality or desire—by talking about how their bodies look. They will say something like 'I felt like I looked good'" (Tolman, qtd. in Orenstein 172). In other words, the risk of socializing toys that encourage girls to perform not only thematic content but also their identity as girls is encouraging girls to see themselves from the outside. Just as the girl in the advertisement enacted her empowerment feminist version of creativity as a theatrical performance for her mother, gendered LEGO suggests that the construction of the self in creative play is, at least for girls, learning to perform the socially-constructed scripts that define girlhood.

This bifurcation of the LEGO medium through its gendered scripting and marketing threatens the fundamental identity of the medium itself. Consequently, it threatens the performances of identity that this expressive medium is designed to facilitate. To be a medium of bricolage is to be a medium open to new permutations and combinations, a medium that derives meaning from creative connections. Directly contradicting this, the implicit playscripts gendering LEGO suggests that the LEGO medium consists of products that are not to be mixed (reinforcing a segregationist logic often found in LEGO marketing). Moreover, LEGO genders not only its products but also its modes of play, masculinizing construction and feminizing dramatization, thereby denying the constitutive interplay that I would argue constitutes a primary force underlying the potential profundity of LEGO.

Post-script 3—Domesticating the Death Star with Steve Price's *The Friends Star*

When targeted marketing aims at segmenting a consumer base, its messages may be taken quite differently by consumers who fall outside the primary targets. Such is the case for AFOL (Adult Fan of LEGO) builders who, as bricoleurs, often scrounge elements from a wide variety of sets not directly targeted toward them. Situated in a perfect position to draw from and recombine many LEGO lines to produce new remixed messages, AFOL builders can literally make something different of LEGO than otherwise intended. As Jonathan Bender explains, "I've developed a bad habit of stealing pieces from sets before I've even built any of the creations. I still read through the instructions, looking at building techniques and the elements mix, but I'm now looking at sets as raw[22] materials rather than as finished products" (2010, p. 219). From this remix-oriented perspective, LEGO Friends can be interpreted as more than a new binarized choice—instead, it functions as a kind of Cambrian explosion that significantly diversified the LEGO palette. As this infusion of new elements was organized around a stereotypically feminized aesthetic, it facilitated representational possibilities largely unprecedented within the LEGO medium. Consequently, the infusion of signifiers designed to script and binarized gendered play simultaneously constitutes an infusion of remixable elements that facilitate transformative play. In LEGO bricolage, the very elements that gender play also allow playing with gender.

In practice, this playful bricolage can mean many things, especially given the complex gender dynamics at play in this predominantly male community. Sometimes, it means that AFOL builders appropriate Friends elements as gender-neutral atoms of shape and color as, for instance, the expanded palette can be used to produce more realistic mosaics or

[22] The rhetoric of 'raw' material subtly implies that when Bender (2010) and other builders see the potential beneath the finished product they are directly perceiving an inherent abstraction in the medium rather than engaging in the active cognitive process of abstracting away from an inherent significant. Analogous to Lisa Gitelman's (2016) argument that *raw data is an oxymoron*, I would argue that the idea of LEGO as raw material represents a culturally-constructed stance toward LEGO that can be useful in practice but is misleading in theory.

architectural models.[23] And sometimes, the fan creations can extend or reinforce the problematic messages of the Friends line.[24] Yet, fan creations frequently also perform transformative play that remixes and potentially challenges the gendered messages behind LEGO Friends. Often the surprising or comic juxtaposition of gendered elements challenges the binary opposition established by targeted marketing, restoring the playful remixing that defines the LEGO medium as a medium of bricolage. To illustrate this transformative potential, this Post-Script explores how Steve Price's *The Friends Star* playfully reimagines the popular LEGO Death Star (Set #10188; see Wolf 2014), domesticating and thereby *defamiliarizing*[25] its core military-industrial logic.

Whereas many LEGO Friends *Star Wars* crossover MOCs (My Own Creations) end up militarizing Friends, *The Friends Star* does so in ways that also narratively and thematically demilitarize *Star Wars*, as the Friends acquire the Death Star on surplus in order to "To Spread A Little Love And Happiness Throughout The Universe And Give It The Ultimate Makeover!!!" (2015, n.p.) Somewhat reminiscent of the aggressively nonviolent "flower power" movements of the 1960s, *The Friends Star* militarizes domesticity to take a proactive anti-war stance. Thus, rather than disarm this battle station, the Friends redesign domesticates militarization by transforming its central weapon into a Love Ray, a weapon for spreading peace throughout the galaxy. Although the narrative of feminized domesticity as an antidote to masculinized militarism is neither new nor particularly feminist, the primary creative intervention here is the integration of the domestic and affective symbology into the war (love?) machine itself. Here, peace is not the absence of war, but rather an active and organized campaign that is technologized, tactical, and heroic in all the same ways that military campaigns are often presented. Rather than

[23] LEGO itself practiced a similar crossover in introducing some new colors in the modular building series that became common in LEGO Friends.

[24] Taking the abstraction of the bricoleur to an implausible extreme, the rhetoric of some fans suggested that the possibility of creating clever pink-and-purple spaceships meant that the elements could be fully separated from their socializing messages (an argument that, in my opinion, falls apart because it hinges on these spaceships being recognizably feminized).

[25] I allude very generally here to the idea of 'defamiliarization' or 'estrangement' introduced by Russian formalist Viktor Shklovsky. Making the familiar seem unfamiliar, as these uncanny militarized Friends builds do, is one way art can question deeply ingrained cultural assumptions.

abandon the Death Star's militarized design, therefore, *The Friends Star* thoroughly interweaves militarization and domesticity to such a shocking degree that it defamiliarizes and thereby undermines commonly gendered assumptions about the relationship between domesticity and war.

As defamiliarization is predicated on making the familiar unfamiliar, Price explicitly plays *The Friends Star* off the well-known Death Star set, whose rooms depict well-known scenes from the films. One telling inversion is found in the Juice Crusher (Fig. 3.7, lower left), which repurposes the Death Star trash compactor. Reversing the direction of flow, the revitalized compactor allows organic waste to be recycled and sent upwards

Fig. 3.7 Steve Price's *The Friends Star*, featuring the Control Room (upper left), the Party Room (upper right), the Love Ray Room (middle left), Juice Bar (middle right), Juice Crusher (lower left), and Hair Salon (lower right) (Image courtesy of the artist)

to fuel the healthy and fun Juice Bar (Fig. 3.7, middle right). Whereas the Death Star is a symbol of destruction that produces soldiers, war, and waste without fostering life, *The Friends Star* is a symbol of production that produces food, entertainment, and Friends (both literally and metaphorically) without depicting waste. This symbolic inversion, moreover, jarringly mixes the visual iconography of these opposing storylines to defamiliarize the messages of both textual worlds. *Star Wars* is construed as more violent and the Friends as more saccharine through their thematic juxtaposition.

Similarly, defamiliarizing the literal military-industrial complex that is the Death Star battle station, *The Friends Star* highlights the original's conspicuous absence of domestic space by reorienting its spaces around hyper-domestic leisure activities. Unlike well-known science-fiction space stations like *Deep Space 9* and *Babylon 5*, the Death Star is completely devoid of domestic space. The only depicted space that remotely implies inhabitation is Princess Leia's prison cell, a clearly militarized environment. Instead, every depicted space thoroughly functionalized within the military-industrial complex—even the trash compactor is filled with industrial rather than domestic waste. Flipping this script, *The Friends Star* regiments leisure with the same military precision, dividing the domestic space into zones—the Pizza Room, Pamper Room, Music Room, TV Room, and Fluffy Pet Room—that uncannily reflect the industrial logic of mass production by being unnecessarily specialized into and organized. Industrializing everyday domestic activities in this way, I suggest, does not so much humanize the battle station as defamiliarize its dehumanizing structure, calling into question how these regimented spaces instrumentalize their inhabitants to the extent that they become mere cogs in the war machine (for more on the instrumentalization of Stormtroopers, see the Chapter 5, Post-script).

In precisely this vein, the uncanniest vignette in *The Friends Star* is an industrial assembly line of Star Wars battle droids moving through "the Stephanizer" to be transformed into an army of Stephanies. This twist goes beyond demilitarizing the built environment and instead presents *The Friends Star* as a factory for domesticating and demilitarizing agents of war. At the same time, appropriating the visual rhetoric of mass-production eerily dehumanizes the LEGO Friends characters by exposing the fact that their individual personalities are replicated into thousands of identical commercial toys (although the Friends characters are easily duplicated if one purchases multiple sets, multiple instances of a single character would never be displayed side-by-side within the LEGO storyworld). This does not present the domestic as a retreat from the industrial but instead presents domesticity—and even the individual characters of the named Friends characters—as products of an industrial system.

Even as this defamiliarizes the industrial logic already at play in militarized LEGO toys, it also defamiliarizes common assumptions about domesticity. In particular, these bizarre and comic vignettes expose how contemporary domestic ideals are often rooted in the ideology of industrial capitalism,. In this context, feminized labor is culturally constructed as linking smart consumption (domestic economy) with domestic production (affective labor). In *The Friends Star*, this is represented in two pairs of linked retail and leisure spaces. The Juice Bar and Pizza Room represent the poles of food service as retail and leisure discussed in this chapter and the Hair Salon and Pamper Room do something similar for the beauty industry, interestingly situating the feminized body as both the object and subject of beautifying labor. These rituals are particularly exposed as culturally constructed performances as the origin story of the Stephanies as repurposed battle droids deconstructs the feminization of food and beauty work. The process of Stephanization, in other words, entails not only a transformation of physical form but an induction into culturally gendered performative rituals.

Domesticating the battle droids suggests that their role as war machines is an artificial rather than essential identity. Reciprocally, rendering the Friends as robots suggests that their domestic orientation is similarly artificial, an expression of programmatic socialization rather than an essential femaleness. Rather than accepting or even inverting the binary logic which presents standard LEGO and LEGO Friends as both separate and unequal, this calls the binary itself into question. The uncanny and defamiliarizing transformation of battle droids into Friends reveals the ideological underpinnings behind both gendered representations. Furthermore, by giving these rather divergent characters a shared origin within the mass-produced world of LEGO toys, this MOC makes a powerful statement that the masculinized militarized Star Wars battle droids and the feminized, domestic Friends are truly kin.

Sharing an essential identity in LEGO plastics makes these figures constitutionally mixable in ways that belie the paratextual messages that attempt to compartmentalize LEGO play. This remixability illustrates a core tension of fan participation in scripted LEGO play, as fans are simultaneously invited into a medium that suggests remix and messages that script and thereby close off certain potential remixes. In this tension, the remix of *The Friends Star* deploys the binarized significances of LEGO's gendered play themes to defamiliarize them, revealing how the performativity of LEGO's socializing play scripts always offers its users some potential for using the medium to facilitate its self-critique.

Works Cited

Armstrong, Frances. 1996. The dollhouse as ludic space, 1690–1920. *Children's Literature* 24: 23–54.

Baichtal, John, and Joe Meno. 2011. *The cult of LEGO*. China: No Starch Press.

Barthes, Roland. 1972. *Mythologies*, trans. Annette Lavers. New York: Hill and Wang.

Bartneck, Cristoph, Qi Min Ser, Elea Moltchanova, James Smities, and Erin Harrington. 2016. *PLoS One* 11 (5): 1–27.

Bem, Sandra Lipsitz. 1993. *The lenses of gender: Transforming the debate on sexual inequality*. New Haven: Yale University Press.

Bender, Jonathan. 2010. *LEGO: A love story*. Hoboken: Wiley.

Black, Rebecca W., Bill Tomlinson, and Ksenia Korobkova. 2016. Play and identity in gendered LEGO franchises. *International Journal of Play* 5 (1): 64–76.

Butler, Judith. 2008. *Gender trouble: Feminism and the subversion of identity*. New York: Routledge.

Cook, Roy T. 2017. Ninjas, Kobe Bryant, and yellow plastic: The LEGO® minifigure and race. In *LEGO and philosophy*, ed. Roy T. Cook and Sondra Bacharach, 91–101. Hoboken: Wiley Blackwell.

Driscoll, Catherine. 2015. The doll-machine: Dolls, modernism, experience. In *Dolls studies*, ed. Miriam Forman-Brunell and Jennifer Dawn Whitney, 185–206. New York: Peter Lang.

Fleming, Dan. 1996. *Powerplay: Toys as popular culture*. Manchester: Manchester University Press.

Gitelman, Lisa. 2016. "Raw data" is an oxymoron. In *New media, old media*, ed. Wendy Hui Kyong Chun and Anna Watkins Fisher, 167–176. New York: Routledge.

Hains, Rebecca C., and Jennifer W. Shewmaker. 2019. "I just don't really, like, connect to it": How girls negotiate LEGO's gender-marketed toys. In *Cultural studies of LEGO*, ed. Rebecca C. Hains and Sharon R. Mazzarella, 247–269. Cham: Palgrave Macmillan.

Halberstam, Jack. 2018. *Trans*: A quick and quirky account of gender variability*. Oakland: University of California Press.

Hardt, Michael. 1999. Affective labor. *Boundary* 26 (2): 89–100.

Jay, Nancy. 1981. Gender and dichotomy. *Feminist Studies* 7 (1): 38–56.
Johnson, Derek. 2014a. Chicks with bricks: Building creativity across industrial design cultures and gendered construction play. In *LEGO studies*, ed. Mark J.P. Wolf, 81–104. New York: Routledge.
Johnson, Derek. 2014b. Figuring identity: Media licensing and the racialization of LEGO bodies. *International Journal of Cultural Studies* 17 (4): 307–325.
Kline, Stephen. 1993. *Out of the garden: Toys, TV, and children's culture in the age of marketing*. London: Verso.
Landay, Lori. 2014. Myth blocks: How LEGO transmedia configures and remixes mythic structures in the Ninjago and Chima themes. In *LEGO studies*, ed. Mark J.P. Wolf, 55–80. New York: Routledge.
Lee, Jonathan Rey. 2019. Master building and creative vision in *The LEGO Movie*. In *Cultural studies of LEGO*, ed. Rebecca C. Hains and Sharon R. Mazzarella, 149–173. Cham: Palgrave Macmillan.
The LEGO Group. 2014. A short presentation. LEGO.com. https://www.lego.com/r/aboutus/-/media/about%20us/media%20assets%20library/company%20profiles/the_lego_group_a%20short%20presentation_2014_english_ed2.pdf. Accessed 7 October 2014.
Merskin, Debra. 2019. Mia had a little lamb: Gender and species stereotypes in LEGO sets. In *Cultural studies of LEGO*, ed. Rebecca C. Hains and Sharon R. Mazzarella, 271–295. Cham: Palgrave Macmillan.
Nipper, Mads. 2012. LEGO Group commentary on attracting more girls to construction play. LEGO.com. https://www.lego.com/en-us/aboutus/news-room/2012/january/lego-group-commentary-on-attracting-more-girls-to-construction-play. Accessed 30 October 2019.
Orenstein, Peggy. 2011. *Cinderella ate my daughter: Dispatches from the front lines of the new girlie-girl culture*. New York: HarperCollins.
Pickett, David. 2012. Part II: Historical perspective on the LEGO gender gap. The Society Pages. https://thesocietypages.org/socimages/2012/05/15/part-ii-historical-perspective-on-the-lego-gender-gap/. Accessed 15 April 2020.
Price, Steve. 2015. The Friends Star. Flickr. https://www.flickr.com/photos/pricey73/16736883714/. Accessed 16 April 2020.
Reich, Stephanie M., Rebecca W. Black, and Tammie Foliaki. 2018. Constructing difference: Lego® set narratives promote stereotypic gender roles and play. *Sex Roles* 79: 285–298.
Sarkeesian, Anita. 2012. LEGO Friends—LEGO & gender Part 1. YouTube. https://www.youtube.com/watch?v=CrmRxGLn0Bk. Accessed 24 April 2020.
Speaight, George. 1969. *The history of the English toy theatre*. Boston: Publishers PLAYS Inc.

Wolf, Mark J.P. 2014. Adapting the Death Star into LEGO: The case of LEGO set #10188. In *LEGO studies*, ed. Mark J.P. Wolf, 15–39. New York: Routledge.

Zinguer, Tamar. 2015. *Architecture in play: Intimations of modernism in architectural toys*. Charlottesville: University of Virginia Press.

CHAPTER 4

Digital Analogs: Bricks, *Worlds*, and *Dimensions*

When LEGO debuted in the 1950s, its bright and clean plastic construction typified the look and feel of modern design, appropriating industrial materials to bring an aesthetic of sleek simplicity to domestic spaces. Somewhat surprisingly, this modern material has also found an affinity with the fragmentary and self-ironizing postmodern logic of the digital age. Noting that "Lego plays a far more important role in the lives of computer people than in the general population," postmodern artist Douglas Coupland writes that LEGO "anticipates a future of pixelated ideas. It is digital" (1995, p. 82). Yet, whereas Coupland sees in LEGO an increasingly technologized future, Mark Frauenfelder sees a digital mindset equally applicable to everyday, non-technologized reality, writing that "With LEGO on the brain, the physical world becomes a hackable platform, and LEGO is the ideal prototype for designing on the fly" (qtd. in Baichtal and Meno 2011, foreword). In this view, ***digitality*** is less dependent on computer technology and more on using digital paradigms—conceptual structures predicated on discreteness or countability—for thinking about the world. Based on the many ways LEGO embraces and explores digital logics, I argue that LEGO is neither digital nor analog—it is both digital ***and*** analog. More precisely, LEGO is a ***digital analog***, a material (analog) system that fundamentally manifests (analogizes) a digital paradigm.

Certainly, contemporary transmedia LEGO maintains a significant presence in so-called ***digital media*** like apps and videogames. In these

J. R. Lee, *Deconstructing LEGO*,
https://doi.org/10.1007/978-3-030-53665-7_4

contexts, LEGO proclaims its status as a digital analog by explicitly embracing the complementarity of its material toys and these environments. As the LEGO Foundation report *Defining Systematic Creativity in the Digital Realm* argues, "The choice between online and offline versions of an activity is a false one: the real opportunity lies in connecting the two" (Ackermann et al. 2009b, p. 40). Contemporary LEGO cannot be understood as either toys or digital media but as a complex web of textual and paratextual threads woven across toys and digital media.

Yet, whereas there is nothing unusual about a contemporary transmedia franchise seeking complementarity across multiple media, LEGO claims an unusually strong material foundation for this complementarity. As the aforementioned report argues, "As the LEGO Group continues on its journey to invent the future of play, the LEGO idea, once simply conceived in the form of a brick, when coupled with the qualities of the digital realm, proves to be a powerful metaphor and tool for the kind of creativity possible in the 21st Century" (p. 85). Although it may seem strange to call a material toy (or its prevailing idea) digital, this report perceptively argues that LEGO is digital precisely because this LEGO *idea* follows in inherently digital logic. Tellingly, when this report argues that LEGO can "invent the future of play" by facilitating "the kind of creativity possible in the 21st Century," it is making a materially-grounded argument for *systematic creativity* already explored in a paired report titled only *Defining Systematic Creativity* (Ackermann et al. 2009a). Tying these reports together, *systematic creativity* is how the reports name the essential material bricolage whose underlying logic LEGO extends into the digital realm.

After all, despite the toy's lack of computational power, it is worth interrogating the digitality of any medium so thoroughly characterized by systematicity, modularity, and bricolage.[1] Together, these reports argue

[1] When discussing digitality and bricolage, it is important to note Sherry Turkle's insightful work on bricolage as a style of computer programming that involves the experiential and experimental manipulation of code in contrast to the 'systematic' approach. She develops this view in *Life on the Screen* (1997) as well as in a collaboration with Seymour Papert (1991) in *Constructionism*. Her use of bricolage differs significantly from mine and, indeed, from many other theoretical uses of 'bricolage,' such as that of the fine arts. Rather than draw upon Lévi-Strauss' (1966) description of the traditional bricoleur, Turkle develops her use of the term from his application of bricolage to contrast the more organic mythology of 'primitive' cultures with the more objective science of 'civilized' cultures. Thus, Turkle sometimes uses bricolage interchangeably with 'tinkering' to

that material and virtual LEGO alike facilitate a postmodern brand of creativity reflective of the systematic digital logic of bricolage (Lee 2019). In other words, whatever else LEGO may materially represent, the medium necessarily toys with digitality itself. As a medium of bricolage, all forms of LEGO play produce digital assemblages of discrete, modular elements: the physical assemblages of construction play, the symbolic assemblages of dramatic play, the narrative assemblages of transmedia play, and the communal assemblages of attachment play.

In Oz-like fashion, digitality turns out to be something LEGO had all along, as the history of LEGO can attest. Like the story of Oz, LEGO only learned to recognize what it already had after wrestling with self-doubt as the rise of other digital media precipitated a crisis for LEGO: "By the late 1990s, there was the distinct sense among many in the LEGO community that with the rise of video games and Nintendo's Game Boy—which by 1998 had sold more than sixty-four million units worldwide—digital play experiences were about the overtake the decidedly physical, tangible experience of building with the brick" (Robertson 2013, p. 80). And this self-doubt led to some questionable decisions—as Robertson describes, this anxiety about competing with 'blue ocean' digital markets precipitated a flurry of ill-conceived digital experiments that nearly precipitated a financial collapse in the early 2000s. Only by reinvesting in their core products was LEGO able to recoup enough financial stability in the late 2000s to conduct a much more successful round of digital experiments.

Moving beyond its initial uneasiness, LEGO eventually embraced the complementarity (and profitability) of digital and analog play as modalities that could reinforce and advertise each other. As Baichtal and Meno note, "A primary reason for the LEGO Group's involvement in the videogame industry is surely to sell its non-software products" (2011, p. 219). Similarly, LEGO confidently claims:

> We consider digitalisation as a way to make our core business – the physical LEGO play – even more attractive and exciting. To do this, we will leverage digitalisation over a broad range of topics – e.g. integrating physical and digital play, e-commerce, marketing and our business/enterprise system. (LEGO 2014, p. 11)

name a process of *discovery learning*. Although this account is quite compelling, it does not directly apply to bricolage as material making. Indeed, the 'structured programming' that Turkle contrasts with tinkering could easily be described as bricolage due to its use of prepackaged subroutines.

Not merely lip service, this philosophy has engendered dozens of more or less successful experiments in interfacing digital and analog bricks.

While many of these simply render the LEGO aesthetic in digital media, some interface between toy and technology in unique ways. For instance, LEGO Mindstorms is a robotics kit that combines programming with the LEGO Technic[2] system, the Life of George (2011–2012) app provides material building challenges that can be scanned and shared by mobile devices with built-in cameras, LEGO Fusion (2014) products populate virtual games with three-dimensional LEGO models extrapolated from brick-built facades placed on special baseplates and scanned with mobile devices, the LEGO Hidden Side (2019–2020) toys feature augmented reality gameplay that links narrative challenges to the physical construction of LEGO sets, and LEGO Super Mario (2020) allows players to construct and play through physical levels that emulate the gameplay of this popular side-scrolling platformer videogame. While not all of these experiments were commercially viable, LEGO's ongoing success amidst the cultural exigencies of the digital era is at least partially due to its bridging analog and digital play.

Given this history, digital play in LEGO is both something it had all along and something that emerged as the brand expanded into new media. Considering digitality as something LEGO has all along, this chapter first theorizes how LEGO reflects a digitality that can bridge material and virtual, building a surprising degree of continuity between the varied products in LEGO's transmedia empire (especially considering the targeted marketing discussed in the previous chapter). Then, considering digital play as an emergent narrative, this chapter examines two case studies of particularly hybrid LEGO videogames—*LEGO Worlds* and *LEGO Dimensions*—that exemplify different aspects of LEGO's digital play. While *Worlds* digitizes LEGO by bringing material LEGO building into the virtual realm, *Dimensions* blends material and virtual play within a transmedia narrative that speaks to the digitality of the LEGO idea. Together, these case studies probe the complex ways LEGO leverages the digitality of its bricolage play to play with digitality itself.

[2] Technic is essentially a sister medium of LEGO, as a construction system that has some traditional LEGO studs but mostly relies on more engineering-like building connections such as pins, gears, and girders.

Digital Analogs

Understanding the relationship between digital and analog is crucial to understanding LEGO, which is at once medium and toy, system and object. As 'digital' and its binary bedfellow 'analog' are loaded and variable terms, this distinction requires some clarification. This section will explore how the digital/analog distinction is used in ordinary language and technical discourse before applying the distinction to the multimedia play contexts in which LEGO functions as a ***digital analog***, that is, a material (analog) system that manifests (analogizes) a digital paradigm.

To show how a material system can manifest a digital paradigm, it is necessary to deconstruct a colloquial use of the term ***digital*** as "Involving or relating to digital or computer technology, esp. the Internet" (OED 2020). This definition of digitality as computation has gained traction for several reasons. First, this relatively recent (1983) distinction ***periodizes*** the digital, allowing scholars to name the epochal social and cultural shifts that accompany the emergence of computer technology. For instance, this helps name what is distinctive about the 'digital age' or 'digital generation.'[3] Second, in a closely related way, this distinction ***technologizes*** the digital, allowing scholars to articulate the cultural impact of technologies. This usage allows 'digital art,' 'digital games,' and 'digital photography' to be contrasted with earlier 'non-digital' forms.

At times it certainly makes sense to group media related by a shared socio-technological landscape. At other times, however, defining digitality as computation obscures the fact that the conceptual relationships exemplified by computational media extend beyond computational media. And this can lead to an overly deterministic picture in which computational technologies are seen as causing these conceptual relationships, when in fact society and technology find expression in each other. Thus, in this project I restore the term 'digital' to its broader sense of conceptually 'discrete or countable' and use other language for computer-aided mediations.

[3] Although such terms turn out to be problematic (for instance, 'digital generation' assumes a generational identity that is undermined by large variations in individual practices or differential access to technology), there is certainly some utility to naming phenomena to promote critical discourse.

Ironically, whereas the colloquial definition of digitality as computation is strongly tied to computer technology, the technical definition of digitality is much less so. In technical terms, the digital and analog distinction can be described as follows:

> 'Digital' simply means that something is divided into discrete, countable units—countable using whatever system one chooses, whether zeroes and ones, decimal numbers, tally marks on a scrap of paper or the fingers (digits) of one's hand—which is where the word 'digital' comes from in the first place. (Cramer 2019, p. 694)

> Conversely 'analog' means that the information has not been chopped up into discrete, countable units but instead consists of one or more signals which vary on a continuous scale, such as a sound wave, a light wave, a magnetic field (for example on an audio tape, but also on a computer hard disk), the flow of electricity in any circuit including a computer chip, or a gradual transition between colors, for example in blended paint. (p. 695)

Notably, these definitions show both how digitality reaches into the material realm via "tally marks" and "fingers" and how analog signals reach into virtual realms when found on "a computer hard disk" or in "a computer chip." Instead of aligning digitality with computation in general, these technical definitions differentiate digital and analog as differing informational *paradigms*.

These paradigms are, in many cases, matters of representation rather than attributes of things. For instance, temperature—a property of the environment—is intrinsically neither digital nor analog but can be represented as either (an analog thermometer measures temperature along a continuous scale, whereas a digital thermometer measures the same phenomenon as a numerical value). Although these distinctions derive from technical discourses, their emphasis on informational paradigms rather than particular technologies is particularly useful for articulating LEGO's digital analogs, material systems "divided into discrete, countable units."

This idea that digital and analog refer to different paradigms for interpreting phenomena as either discrete or continuous, informs what I shall call the *phenomenological*[4] distinction, which is essentially the technical

[4] I primarily use 'phenomenological' in the broader sense of 'experiential.' There are parallels between my analytical method here and the more nuanced philosophies of

distinction applied to perceivers' experiences of given phenomena. Rather than applying to the bowels of the machine (often a black box in everyday circumstances), the phenomenological distinction applies to *surfaces*—facets that are directly interacted with and experienced. In this view, a phenomenon is digital insofar as it affords being engaged as discrete, countable units. Conversely, a phenomenon is analog insofar as it affords being engaged as continuous and holistic. According to this nomenclature, for example, one can describe the digitality of a clock (Fig. 4.1) according to how it displays—and thereby constructs the experience of—time. Emphasizing this rather than its interior technology raises different questions, such as the implications of representing time as a countable sequence of incrementing yet discrete moments rather than as an inexorable and continuous flow. Thus, whereas the colloquial definition is most useful for naming computers as core elements in a shifting socio-technological landscape, the phenomenological distinction is most useful for articulating how processes of categorizing the world construct its significance for us.

LEGO Bricks as Digital Analogs

All these distinctions are useful as interpretive paradigms designed to yield (and name) different types of insight. Yet, in the case of the material play facilitated by LEGO, I find the phenomenological paradigm most useful in characterizing how LEGO play fuses analog and digital logic. Take, for example, the humble LEGO brick

Linguistically, the word 'brick' implies both a solidity that defines the object as a lump of continuous, analog matter and a set of part-whole relations that define the object as designed for digital assemblage. Materially, the LEGO brick simultaneously adopts the analog malleability of plastics and the digital discreteness of atomistic particles (Lee 2014). Such doublings are, as I see it, constitutive features of the LEGO medium, which facilitates the playful fusion of these seemingly opposed modes. And LEGO not only plays with the interrelationship between digital and analog—the LEGO brand identity has *become* a digital analog that functions as a widespread symbol (analogy) for atomism, digitality, and

phenomenologists like Martin Heidegger and Edmund Husserl, who emphasize how the world *as we know it* is constructed by human perceptual, conceptual, and discursive paradigms.

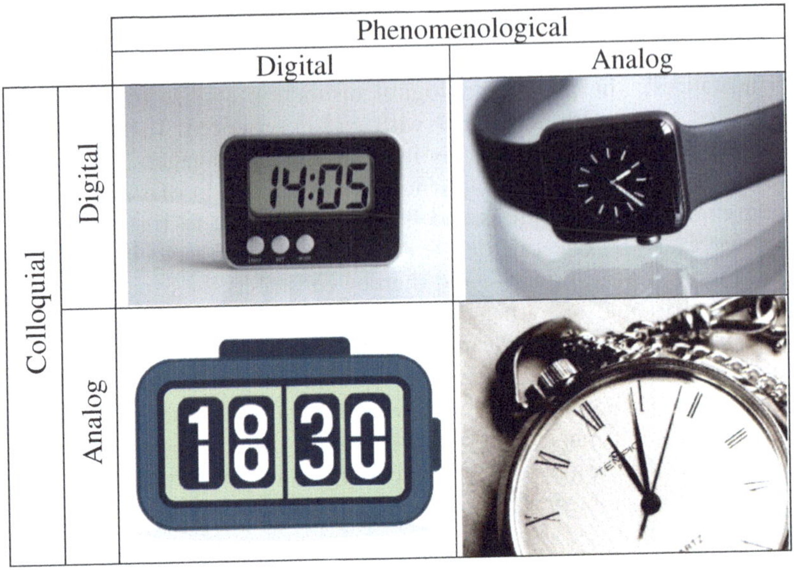

Fig. 4.1 Four timepieces illustrating the disjunction between the colloquial distinction that distinguishes the presence or absence underlying computer technology and the phenomenological distinction that distinguishes between experienced informational paradigms. Note that the bottom two clocks are presumed to operate based on non-computational technology although this cannot be fully determined from the images. The bottom-left drawing is of a *flip clock*, which uses a motor to rotate each wheel at a different speed, causing the leaves to periodically flip, revealing the next number. Even if the motor is electric (as opposed to wind-up), the motor is only needed to supply continuous (analog) motion and needs no computational component

connectivity. In this sense, LEGO not only participates in but represents digital thinking, materializing a digital worldview.

Material LEGO toys are digital analogs in a variety of ways, but especially in their *collectability* and *connectivity*. Collectability has long been a defining LEGO ethos, appearing in the foundational LEGO principle "Extra sets available" (Baichtal and Meno 2011, p. 14). LEGO employs many strategies to encourage its existing consumers to perpetually expand their collections: organizing sets into lines of uniquely numbered yet complementary products, releasing limited-edition and collectible minifigures, and disseminating checklists,[5] an entirely binary form of record-keeping that allow consumers to record and gamify[6] their collections. After all, "That need to acquire more pieces is easy to understand," as Jonathan Bender notes, "A LEGO collection is quantifiable. You can physically see how many bricks you own" (2010, p. 19).

This completionist impulse is further reinforced by LEGO's media narratives, which invariably revolve around *collectives* (tight-knit groups like the LEGO Friends or Ninjago ninja) and *collections* (in many plots,[7] the central quest narrative involves the collection of a set of artifacts). Collection is also a central mechanism for various LEGO videogame series,[8] even though virtual objects are theoretically unlimited. All the adventure videogames mentioned in this book exhibit two types of collectability: a currency of 'studs' that can be accumulated through the course of the game and unlockable content, especially minifigures, that are often explicitly presented as collectibles.[9] In all these ways, collectability reinforces a phenomenologically digital logic, a way of perceiving one's relationship to material property according to binary (ownership vs. non-ownership) set logics.

Collection, that is, replaces scrounging as the first stage of LEGO bricolage. While all forms of bricolage creatively reassemble elements drawn from a reservoir of objects, LEGO's consumerist, brand of bricolage constructs this reservoir from collectible rather than found objects.

[5] Mattes (2019) interrogates how LEGO's play with lists and listing operates in a *post-ironic* mode.

[6] *Gamification* names the practice of using game mechanisms—most commonly reward systems—for motivating everyday activities. Here, the satisfaction of completely checking off all the items in a checklist extrinsically motivates collection (as opposed to the intrinsic motivation of deriving utility or pleasure from the items being collected themselves).

[7] Although this is a far from exhaustive list, here are a few play themes whose plot involved collecting various artifacts: Bionicle, Elves, Legends of Chima, LEGO *Star Wars*: The Freemaker Adventures, LEGO *Star Wars: The Yoda Chronicles*, Ninjago, Power Miners, and Rock Raiders. In addition, there are multiple play themes that have relatively little narrative but imply the activity of collecting some kind of resource, such as the treasures in Aqua Raiders and Pirates or the ice crystals in Ice Planet 2002.

[8] For more on collecting digital objects in videogames, see Toups et al. (2016).

[9] As Robert Buerkle writes, "Since the story is already known, and the levels fairly easy to finish, one could argue that the *real* goal of the game is not so much completing the story as it is in completing your collection—after all, the reward for finishing each level is in being shown the minifigs that you've unlocked, as well as the number of other collectables that you still have left to mark off your list" (2014, p. 141).

Curated consumption precedes creative reassembly. As Susan Stewart notes, collection is shaped by a constitutive paradox in which "we are inheritors, not producers, of value" (1984, p. 164) and yet "Not simply a consumer of the objects that fill the décor, the self generates a fantasy in which it becomes producer of those objects, a producer by arrangement and manipulation" (p. 158). As consumption that generates a fantasy of production, collectible LEGO anticipates the bricolage it makes possible, setting the stage for the participatory medium of LEGO to more thoroughly blend consumption and production.

Thus, collection sustains LEGO's *connective* potential, which is directly proportional to its available part set. Whereas the economic principle of *diminishing marginal utility* proposes that the more of something one acquires, the less additional value it provides, LEGO promises to compound value based on the counter-intuitive principle "The more LEGO, the greater the value" (Baichtal and Meno 2011, p. 14).[10] While only a few pieces are needed to provide functionally limitless combinations—just six 2 × 4 bricks of the same color offering 915,103,765 combinations[11]—many more pieces are needed to build more complex structures. As LEGO elements become interesting only when embedded within this combinatoric system, quantifiable collections provide the digital possibility space for LEGO construction play.

The connective potential of LEGO, moreover, is itself digital in a strikingly tangible way. LEGO elements are designed according to mathematical ratios that ensure that they can connect into an implied Cartesian grid. Consequently, these *atomistic* elements interact not only with each other but also with a space of possibility that has a fundamentally digital logic—like binary bits, LEGO bricks are either tangibly present or conspicuously absent within the regularity of the grid. The pixelated aesthetic of LEGO visually represents the pixelation of the material system itself, distinctively analogizing digitality. Thus, as collection sustains the connective potential of LEGO, connection organizes the collection. The digitality of the LEGO commodity and the digitality of LEGO construction play go hand in hand.

[10] Technically, this phrase would also apply to commodities with diminishing marginal utility if it is interpreted as "the more X, the greater the *total* value." It is clear, however, that LEGO aims to produce a systematic, synergistic value that goes beyond this.

[11] See *A LEGO Brickumentary* (2014).

Although this logic of collection is predicated on the immanent scarcity of finite objects/commodities, it also permeates computer-mediated virtual environments that have no such finitude. Rather than transcend the material conditions of the brick to fully embrace virtuality, LEGO consistently performs dialogues between material and virtual play. Thus, while material LEGO is already a phenomenological fusion of digital and analog, this fusion only becomes more complex as LEGO continues to expand its videogame, online, and app content. Through both collectability and connectivity, the core messages of LEGO play make an already digital medium even more so—encouraging consumers to firstly count their collection and secondly follow the digital gridlines when building with said collection. LEGO actively promotes its own brand identity according to the ways it transcends this potential schism, situating itself as a toy medium and media toy for an age of postmodern play.

Making tangible a digital way of thinking, LEGO is a medium whose message is, at least in part, how divides up meaningful elements. Conveying its messages through discrete systems of symbolic elements, LEGO is a phenomenologically digital medium. And yet, LEGO is also a tangible construction toy. Just as digital conceptuality is rooted in the continuous-yet-dividable fingers of the hand, LEGO is a material object system that formally reflects digital conceptuality—LEGO is a digital analog. Thus, LEGO play involves playing with digitality itself, pushing the boundaries between the material and the virtual, analog and digital. Such play itself is a digital analog insofar as its unit operational engagement with the LEGO toy or game is bound up with an implicit critical exploration of the digitality of matter (i.e. atomism), computers (i.e. code), and even conceptuality itself.

Building on this conception of material LEGO as a digital analog, the following sections explore how the virtual implementations of *LEGO Worlds* and *LEGO Dimensions* explicitly probe the intersection of material and virtual, digital and analog. While they do so in markedly different ways—I characterize the LEGO of *Worlds* as *digitized* and the LEGO of *Dimensions* as *digital*—both play with mediating the inherent digitality of the LEGO medium and ideal by refracting it through virtual, computational gameplay that itself mixes phenomenologically digital and analog play experiences.

Digitizing Toys in *LEGO Worlds*

Beyond the colloquial definition of 'digitality' as 'computation' lies another, narrower definition of 'digital' as "Designating a virtual, computer-mediated counterpart of an object that exists in the physical world" (OED 2020). This refers to *digitization*, the process of computer-aided conversion into bits and bytes. To digitize a toy is to render its material play int(e)ractable within a virtual environment. Like other translations or adaptations, digitization raises questions of similarity and difference as the 'same' core elements are re-presented in phenomenologically distinct ways. This play of identity and difference manifests most significantly in digitization as a form of *remediation*, the process by which new media reconfigure (rather than merely replace) prior media forms. After all, digitizing LEGO requires more than visually representing the LEGO aesthetic the way LEGO merchandise might bear images of LEGO bricks. Instead, digitizing the LEGO medium entails crafting a virtual analog to the *mediateness* of material bricks, namely their distinctive int(e)ractive potential for mediating ideas through bricolage.

This section conducts a case study of the 2015 sandbox construction videogame *LEGO Worlds*[12] to consider how the process of digitization simultaneously reframes and transforms the inherent digitality of the LEGO medium. Although *Worlds* is not LEGO's first attempt at digitizing its material toys,[13] this digitization is particularly interesting because it combines virtual construction play reminiscent of LEGO Digital Designer,[14] free design software for designing LEGO creations, with gameplay elements reminiscent of LEGO adventure games (see Chapter 5). *Worlds*, that is, is a hybrid of game genres similar to the

[12] Since its prerelease on Steam in 2015, *LEGO Worlds* has undergone a number of updates and patches that introduce variation into this game. The patch referenced in this chapter was released on Steam on September 13, 2018.

[13] In 2010, LEGO released a massively-multiplayer construction game *LEGO Universe*, which was shut down less than two years later.

[14] Presented as niche software for hardcore fans rather than as a game, LEGO Digital Designer has been slowly phased out as Design byMe (a program that allowed builders to order custom sets based on their designs) closed in 2012 and the company has since stopped supporting the software.

popular sandbox building/adventure game *Minecraft*, with which LEGO has a longstanding and complex relationship.[15]

In translating the physical immediacy of material LEGO toys into virtual environments, *Worlds* constructs a distinctive interface that redefines the mediateness of LEGO. Algorithmically rendering material bricks as digital pixels, the hybrid play in *Worlds* offers a virtual analog to the digital analog of LEGO bricks that probes the fundamentally digital mediateness of LEGO construction.

The first way *Worlds* digitizes LEGO is by creating a three-dimensional environment that functions as a virtual analog to the inhabitable spaces of material play. As is common in videogames of this type, this game relies upon a pair of reciprocal material analogies: (1) the two-dimensional screen projecting a view analogous to a window into a three-dimensional world, and (2) game inputs projecting the real movement of players into analogous onscreen movement. These material analogies function as Mattia Thibault calls *digital prostheses*, "virtual equivalents to analog play, allowing players to interact with virtual playthings as they normally do with real ones" (2019, p. 243). Focalizing these mediating projections, collectible and controllable minifigure avatars become "invested with the role of *actant observer*" (p. 243), organizing players' agential and perspectival engagement with the virtual worlds. As players move these avatars around the virtual world, discovering locations, collecting items, and fulfilling tasks for non-player characters, these int(e)ractable analogies of the game window and game inputs mediate a real, material, and phenomenologically analog experience. As the game window amplifies the gaze and the game inputs amplify player movement, the game constructs phenomenologically *analog* experience of material play.

[15] *Minecraft* designer Markus "Notch" Persson openly acknowledged his love of LEGO as an inspiration for his cube-based building game (Goldberg and Larsson 2013, pp. 28–29). This similarity was not lost on consumers, and many comparisons between the two were made in popular media and discourse well before LEGO and *Minecraft* developed a formal licensing arrangement in 2012, when a *LEGO Minecraft* set (Set #21102) was selected as one of the user-designed sets in the *LEGO Ideas*—then *Cuusoo*—product line. Although *LEGO Worlds* is not quite a LEGO-skinned version of *Minecraft*, it fits the general description of the game described in the booklet that accompanied Set #21102 as "a sandbox construction game" which "involves players creating and destroying various types of blocks in a three-dimensional environment. The player takes an avatar that can destroy or create blocks, forming fantastic structures, creations and artwork."

Calling this virtual play analog might seem strange if one applies the colloquial definition of digitality as computation. Yet, this surprising reversal is quite common in game design, as the so-called 'digital' medium of videogames increasingly aims to emulate phenomenologically analog experience by using improving technologies (processors, displays, haptics, touch and motion-sensitive inputs, etc.).[16] Similarly, a primary selling point of the aptly-named *LEGO Worlds* is the promise of exploring richly-detailed and whimsically-designed LEGO worlds. Each of the game's more than two dozen biomes features a distinctive brick-built landscape populated with creatures, vehicles, and sometimes interactive animations that lend thematic flair to otherwise simplistic gameplay. Whereas many games (including most LEGO adventure games) use three-dimensional landscapes as mere backdrops for gameplay with other selling points—combat, puzzles, narrative, etc.—a primary purpose of *Worlds* is to explore and interact with a LEGO landscape that digitizes the experience of playing with LEGO toys.

In particular, this three-dimensional exploratory play intensifies the *miniature sublime* (see Chapter 2) by analogizing how miniature LEGO toys play with worlds too small to inhabit. Compared with material toy play, the virtual projection that translates player input into onscreen avatar movement adds a layer of mediation that simultaneously insulates players from the virtual world while giving players more control over how toys move through this inaccessible world. This intensifies the paradox of the miniature sublime—the simultaneous feeling of being invited to participate in a world and feeling of being excluded from that world. As material LEGO toys are readily accessible yet too small to inhabit, the immersive virtual spaces of *LEGO Worlds* nostalgically remediate the miniature experience while promising to take players further into LEGO worlds than they have ever gone before. In this push and pull, the digitized virtual spaces reflect the poignant contradictoriness of two media—toys and videogames—that simultaneously invite and exclude.

How the Adventure mode remediates the sublimity of an analog toy experience, however, is only half the story. Whereas Adventure mode (Fig. 4.2a) disguises the essential mediateness of the videogame

[16] Conversely, games built on analog technologies (i.e. board games) maintain a phenomenologically digital substructure by relying on discretized textual elements like game boards, tiles, cards, dice, currency, counters, and pawns that maintain a digital conceptual structure.

Fig. 4.2 Two screenshots comparing **a** the Adventure mode and **b** Free Build mode in *LEGO Worlds* (9/13/2018 software update on Steam). The Adventure mode shows the minifigure character in the midst of an animation, the health in the upper-left, and the map in the upper-right. The Free Build mode shows the same character in the lower-left corner building the word "LEGO" using the Build Tool. The screen is automatically centered on the selection tool, which shows a transparent green rectangular volume on the right side of the letter 'O' being used to select an area to fill with bricks (note the "Drag" command displayed in the center). Releasing the "A" button (This interface and its depicted inputs would slightly change if the game was set to a different controller, such as keyboard/mouse.) at this moment would complete the letter. Other tools and options are displayed in the bar along the top

to promote immersion, the digitized LEGO construction in Free Build mode (Fig. 4.2b) overtly remediates the media-specificity of a construction toy system. To accomplish this, this mode first suspends the focalization of the avatar, which becomes fixed in space and decentered in the frame. As the avatar gives way, gameplay instead becomes primarily mediated by a virtual game interface featuring visible commands, tools, cursor, and gridlines. Contrasting with the more streamlined, immersive interface of avatar-centric Adventure mode, these more explicit visual cues emphasize the media-specificity of the game system. This facilitates the transition from the phenomenologically analog experience of digitized LEGO miniatures to the phenomenologically digital experience of digitized LEGO construction play. Here, a parallel emerges between the digitality of LEGO bricolage and the complex array of menus and commands needed for the videogame player to virtually remediate LEGO construction.

More specifically, this mode emphasizes the mediateness of gameplay by interjecting a layer of tool use between player and world. This is a sharp contrast to material LEGO toys, which are designed to be easily constructible by hand—rather than a construction material in need of tools, it makes more sense to say that LEGO is its own construction tool (the one LEGO tool, the brick separator, is more of a deconstruction than a construction tool).[17] Whereas material LEGO facilitates toolless construction, building in *Worlds* requires players to select one of several tools that together constitute the Free Build mode: the landscape tool, discovery tool, copy tool, paint tool, and build tool. Each tool brings up a virtual overlay that provides a unique way of manipulating the brick-built world. For instance, players may select swaths of landscape to flatten or raise, select geometrical areas of space to fill or empty (Fig. 4.2b), select areas or objects to copy-and-paste, recolor existing bricks, and even build "one brick at a time—the old-fashioned way," as the game narrator describes. Built into the game interface, these tools simultaneously structure and visibly mediate digitized construction play. As these tools function as virtual prostheses, they create an experience that is more viscerally about building with tools—especially computer-aided design tools—than building with LEGO.

[17] Elsewhere (2019), I analyze how LEGO Foundation research reports describe the importance of tool use for fostering creativity.

Tools, of course, often symbolize construction. Yet these virtual tools do not symbolize toolless LEGO construction but rather the tool-like mediation of the videogame interface. Thwarting a symbolic connection to material construction, the minifigure avatars neither click pieces together like human hands nor hammer bricks together like typical construction tools. Instead, the tools in the game are futuristic contraptions that perform action at a distance, making material causes and effects the product of a mysterious technological force rather than direct physical intervention. These tools, therefore, operate as virtual analogs to mediated videogame play—after all, videogames also allow players to perform actions across the distance of the mediating screen. Consequently, the in-game building tools analogize videogame play as the space of direct int(e)raction that mediates digitized construction play.

Furthermore, the emphasis on tool-mediated construction is accompanied by a complete deconstruction of the constraints and affordances of material LEGO. The entire possibility space of this virtual construction play is determined by the game tools, representing a radical transformation in the mechanics of construction play. Here are a few notable examples:

- *The suspension of physical laws.* In *Worlds*, bricks may be placed anywhere in space, where they remain fixed and unaffected by gravity. Also, bricks can be placed intersecting existing bricks (upon which the computer will automatically adjust the construction to fit the new element). As computational systems are more than capable of simulating basic real-world physics, here the game offers a ludic experience of transgressing typical material constraints.
- *The suspension of LEGO connectivity.* As a consequence of the suspension of physical laws, virtual LEGO bricks no longer need the internal connections that define the medium in order to hang together. As LEGO studs become vestigial, LEGO bricks become reduced to mere solids, particles of pure form more like traditional building blocks than connectible LEGO. This is complicated, however, by the fact that connective patterns that are lost to the elements themselves are reencoded within the constraints of the videogame interface, which retains LEGO-like connective rules.
- *The suspension of analog 'play.'* Although material LEGO is designed around an embodied Cartesian coordinate system, it also facilitates a wide range of analog subversions. For instance, there is enough

'play' for a sufficiently long wall of 1 × 2 bricks to be bent into a complete circle. As algorithms force virtual LEGO elements to 'snap to grid,' they lose all possibility for continuous ranges of motion or articulation that define the phenomenologically analog experience of material objects.

- *The suspension of scarcity.* As Keven Schut writes, "because the digital medium has unlimited amounts of space and virtual blocks, the only restriction for LEGO players is now the amount of time they have" (2014, p. 228). Consequently, the game transforms not only the logic of the construction toy but also the logic of the collectible commodity. As time becomes the central currency in this world, buildable LEGO elements are collected not by spending money but by playing the adventure.[18]

In all these ways, the game remediates the int(e)ractability of its digitized LEGO toys, which are interact-able only within the specific constraints and affordances of the game interface and intractable outside this interface. Thus, the instant the player leaves the Free Build mode and reenters Adventure mode, the buildable LEGO bricks revert to being fixed parts of the landscape.

Whereas many LEGO adventure videogames continually reduce LEGO to merely part of the landscape, *LEGO Worlds* retains an emphasis on construction play. Emphasizing the mediateness of gameplay, this game becomes a virtual analog to material construction play that operates under a transformative analogy—rather than mimic how LEGO toys actually work, the game interface constructs a parallel experience reminiscent of material LEGO construction while being in reality quite different. It does this primarily by reinforcing and even extending the essential digitality of LEGO construction. In particular, as construction without material constraints is functionally equivalent to design and as the game window more closely resembles a CAD interface like LEGO Digital Designer than a traditional adventure game, Free Build mode construes LEGO construction play as an act of *design*. This transformation amplifies how LEGO has always embodied design ideas within its

[18] Complicating this, money can be used to expand the adventure by purchasing DLC (downloadable content) that give players more world to explore.

material forms. For instance, by algorithmically fixing LEGO construction to its embedded Cartesian coordinate grid, virtual LEGO eliminates possibilities that already ran counter to the implicitly structured LEGO design. One might say that this game departs from the materiality of LEGO construction in order to distill LEGO construction to its essential digital idea.

Be it through its pixelated LEGO aesthetic, the rigid Cartesian grid that underlies its atomistic building practices, or the logic of collection that rewards exploration play, *Worlds* provides a play experience that is itself a bricolage of the digital logics of LEGO. Although *Worlds* appears simply digitize an analog medium, it actually mediates an ongoing interplay between digital and analog that characterizes the material LEGO it represents. This game, therefore, presents virtual LEGO as complementary to material LEGO, offering to provide additional dimensions to the already hybrid LEGO experience. This coalesces into a playscript that characterizes both material and virtual LEGO play as the *systematic creativity* described in the LEGO-sponsored research reports. Maintaining the ideologies of "hard fun" and *development, imagination, creativity* that define the LEGO brand (see Preface), *Worlds* digitizes construction play in ways that reflect and algorithmically reify the relationship between mechanical and ideological creative bricolage that defines the material LEGO brick.

Digital Toys in *LEGO Dimensions*

While the doubled mediation of digitized toys in *LEGO Worlds*, develops a parallelism between material and virtual worlds, the 2015 *LEGO Dimensions*[19] (Fig. 4.3) technologically integrates toy and videogame to literally *connect* material and virtual, using a specialized 'toy pad' to translate microchip-embedded toys into videogame inputs. *Dimensions* thus enters the inherently hybrid genre of the *toys-to-life* videogame, a genre that reveals a complex interplay of material and virtual. Looking at *Dimensions*' narrative, advertising, and gameplay will therefore reveal the emergence of LEGO as a digital analog in which material and virtual LEGO work together to construct int(e)ractable analogies of digitality itself. Thus, as the LEGO brand expands beyond the material toy, it emerges as

[19] Like many LEGO games, *LEGO Dimensions* is available on multiple platforms. This analysis refers to the Xbox version.

Fig. 4.3 Product packaging for part of the *LEGO Dimensions* Starter Pack (set numbers vary by game platform) containing the toy pad, minifigures, and LEGO elements. While the overall graphic design of this package closely resembles a typical LEGO set, the brick-built elements fade into the background while the dynamic action is conveyed by the computer-animated videogame characters leaping toward the viewer from a solar system of interdimensional bubbles. This reflects the intermingling of virtual and analog worlds in which *Dimensions* uses the portal to simultaneously promise to transport players into the virtual LEGO world and transport animated LEGO characters into the space of material play (as happens in the online trailers)

a distinctively hybrid form that often leverages analog experience to play with an underlying digital idea.

In this way, LEGO exemplifies the increasing hybridization of a media culture equally obsessed with remediating old media and pioneering new ones. In this vein, children's media scholar Meredith Bak points out that

"Toys that bring 'old' and 'new' media together both satisfy nostalgic, conservative critiques of children's media, which tend to privilege the physical and the tactile over the virtual, and harness the unique qualities of new media" (2016, p. 58). Like its predecessor *Disney Infinity*, Bak argues, *LEGO Dimensions* capitalizes on its brand association with creativity such that the legacy of its material toys inflects the structuring ideology of its videogame play.

The narrative of *Dimensions* frames this hybrid play experience by bringing together an unlikely bricolage of characters from various transmedia franchises[20] in a quest to collect[21] special "foundation elements" needed to save the various LEGO dimensions. Using interdimensional portals to connect these disconnected dimensions (ironically, in order to prevent the supervillain Lord Vortech from connecting them), the game plays with the simultaneous unity and disunity of dimensions tied to different media franchises.

Although this story is predominantly linear, the simultaneity of these multiple worlds analogizes the hybrid form of the game itself. Contrasting the sense of separateness connoted by 'worlds,' 'dimensions' evokes the image of different aspects or facets of a single reality, reflecting the principle of *one reality* defined by the *Defining Systematic Creativity in the Digital Realm* report as "[physical and virtual worlds] do not exist separately from one another, but are increasingly one and the same with play experiences which evolve for traversing back and forth between both

[20] Such toys-to-life games are only one instance of a broader cross-media hybridization that increasingly characterizes our participatory media culture. In addition to bridging toy and game play, *Dimensions* also plays with the intersection of transmedia franchise. As fan studies scholar Matt Hills argues, *Dimensions* undermines its promise "to go beyond the corporate norms of co-branding, bringing many franchises into creative dialogue" (2016, p. 12) due to how, despite "Co-opting transgressive play, such crossover IP mashup remains entirely restricted by corporate ownership" (p. 14). While *transmedia play* is beyond the scope of this chapter and is the primary focus of the next, the fact that the discrete dimensions of this game are tied to licensed transmedia franchises does add another layer of meaning to the media-specific hybridization of the toys-to-life game.

[21] The theme of collection anchors many of LEGO's material play themes, which often situate collectible sets within such a quest narrative. Like many character toys and action figures (Kline 1993, pp. 239–240), such sets also align collection with agency such that the acquisition of specific items confers specific abilities. In this way *Dimensions* differs from the narrative logic of most other LEGO adventure videogames, which typically derive their narrative progression from the teleology of their adapted media texts, and instead adopts the narrative strategy of LEGO's material product lines.

worlds" (Ackermann et al. 2009b, p. 77). Thus, the narrative of *Dimensions* is significant not only because it reflects how "today's children mix and match media rather than start from scratch" (p. 13), but also because it begins to erode the distinction between material and virtual play.

The story begins, as many videogames do, with a tutorial level that introduces the narrative and gameplay. In this level, players must discover piles of virtual bricks in order to repair the destroyed interdimensional portal. When the player issues the command to build these bricks within the game, the diegetic gameplay is interrupted as the screen cuts to a virtual LEGO instruction book that prompts the player to physically construct the portal from the included LEGO elements. This establishes the paradigm of material and virtual connection by having players construct the portal in parallel across the material and virtual dimensions.

According to this paradigm, the building and rebuilding is reciprocally linked to in-game diegetic progress. Rather than build everything at once, the game cues material building linked with story-advancing events (for example, upgrading the 3-in-1 miniature vehicles prompts onscreen instructions for reconstructing the models, which are then reflected in an updated appearance within the game). Like the LEGO instruction manuals embedded within this virtual gameplay, this tutorial structure scripts players' navigation between material and virtual worlds, walking players through the mechanical and symbolic specifics of the game's constitutive material/virtual interface.

Reinforcing this hybrid gameplay, the game's paratextual marketing clearly establishes material/virtual hybridization as *Dimensions*' selling point. As the text in one product video reads, "More than a toy … More than a game … More value" (LEGO 2015). Promising an excess of value resulting from a synthesis that transcends either medium alone, LEGO advertises its hybrid media platform as fulfilling a fantasy of play that is satisfying unto itself. This fantasy is clearly demonstrated by the mixing of realities and play modes that characterizes a pair of online trailers (Warner Bros. 2015a, b) featuring actors Joel McHale (who voices the character X-PO in the game) and Alison Brie (who voices Unikitty in *The LEGO Movie*). Interestingly, neither trailer shows a game interface, instead depicting the virtual world intervening directly in live-action scenes.

In both trailers, constructing a mysterious LEGO portal triggers a collision of material and virtual: McHale is confronted by minifigures who eventually enlist his help rebuilding a damaged vehicle, and Brie—after comically revealing a backstage persona clearly derived from her role as Unikitty—becomes entranced by the light of the portal, implicitly transported directly into the world as the trailer cuts to a montage of scenes from the game. In other words, McHale enacts an analogy to gameplay in which his material assistance contributes to the narrative success of the characters while Brie enacts a fantasy of immersion in which she first becomes her character and second falls into the virtual space.[22] Yet, treating these moments as mystical intrusions upon an ever-present reality breaks such immersion. Instead, *Dimensions* offers a non-immersive, self-reflexive fantasy that derives its ludic pleasure from the simultaneous awareness of material and virtual to fulfill its promise of bringing a new dynamism to imaginative toy play and a new materiality to virtual gameplay.

This dynamism produces a kind of paradoxically deconstructed *augmented reality*. Whereas LEGO Hidden Side sets offer a more literal version of augmented reality in which virtual images are overlaid atop a mobile device's camera-eye view of the material world, the virtual world of *Dimensions* metaphorically augments reality as a cybernetic extension of material LEGO play literalized by a toy pad which allows the manipulation of minifigures to directly impact the game. The toy pad has three areas upon which up to seven different characters or vehicles can be placed at a time. While only three figures and one vehicle are included in the base game, dozens more characters and vehicles are available.[23] To ensure that these toys remain active participants in the play experience, the gameplay makes frequent use of the toy pad in the following ways:

- To import characters and vehicles into your active party.
- To unlock additional DLC (downloadable content).
- To 'break free' from binding attacks by quickly moving characters from the targeted areas before they take damage.

[22] These advertisements implicitly reinforce the gendering of play discussed in Chapter 3, as the McHale trailer emphasizes construction play while the Brie trailer emphasizes dramatic play.

[23] With a few dozen expansions initially available and more potentially forthcoming, these sets take the game well over the price range of a typical videogame. Collecting all the *LEGO Dimensions* sets could cost more than the console the game is played on. For a player to invest so extensively, it is necessary that the LEGO toys themselves contribute significant value to the product.

- To perform 'master builds' by moving characters across areas of the toy pad as they light up sequentially.
- To use the five 'keystones'[24] required to solve level puzzles. Four of these require characters to be moved to different areas of the toy pad (to activate wormholes, unlock obstacles by matching and mixing colors, gain different elemental powers, and change size) while the fifth uses gradients of colored light to indicate proximity to otherwise invisible dimensional rifts.

Despite the simplicity of these tasks, the toy pad is integral to the play experience in two primary ways. First, the binary nature of this input device provides an additional digital element to game controllers that typically mix analog (joysticks/mice) and digital (buttons) inputs. The placement of minifigures on the discrete areas of the toy pad provides a tactile and tactical gameplay more reminiscent of placing pieces on a game board than using standard videogame controllers. Secondly, the toy pad links this play at digitality to the material toys that form the affective locus of the game. As most dolls and action figures mimic action rather sketchily—even the most realistic crying dolls or punching heroes fall rather short from a purely mimetic standpoint—toy play often operates through a symbolism of attenuated *gestures* in which miniaturized motions stand in for more complex actions. An entire imaginative adventure can be elicited by simple swoops of static figures. This is precisely how *Dimensions* derives significance from the relatively simple, digital positional movement of toys on the toy pad. Thus, although such play has nowhere near the material complexity of building or playing with a LEGO set, the use of material toys as game inputs complicates and enriches narrative play.

While the horizontal platform of the toy pad is the literal, technological site of the material-virtual interface, the vertical LEGO portal imaginatively connects these realities. Although the player is situated outside the virtual game world in ways analogous to the miniature sublime, the portal-as-window opens up the virtual world to the player's sensory experience, as the Brie trailer suggests. As a toy, building the brick-built portal establishes a tangible connection between the player and the figurative

[24] These keystones not only anchor the gameplay, but also serve as an important element of the plot (they are collected within the game and subsequently indicated by patterned shield elements which the player then adds to the brick-built portal).

window into the virtual world. Once built, the portal sits atop the toy pad which interfaces toy and game and offers a window into the virtual world. This vantage point is advertised in a product image that literally uses the toy portal as a window through which can be seen a television screen displaying a mirror image of the portal. Indeed, the portal-as-window is a miniature representation of the screen, which is itself a window into the game world.

Yet, whereas screens frame virtual worlds in ways that direct attention through the window to promote immersive visual experiences, the miniature and visually-complex portal draws attention to itself, functioning not only as a transparent window but as a tangible symbol of the material-virtual mediation that screens provide. Drawing attention to the interface encourages awareness of the game as a game and, thereby, opens a space for questioning the relationship between games and toys. The symbolism of the portal, therefore, also undergirds a theme of interconnectivity that both encourages extradiegetic reflection on the nature of its gameplay and structures the diegetic game narrative. Together, the technological mediation of the toy pad and the metaphor of the portal link material and virtual realities to create a gameplay that melds digital and analog play experiences by requiring digital problem-solving in a virtual world that is navigated using a phenomenologically analog interface.

Fittingly, perhaps the most acclaimed dimension in *Dimensions* is the *LEGO Portal* dimension (Fig. 4.4), the first collaboration between LEGO and the popular first-person puzzle/platformer videogame *Portal*, whose innovative gameplay used a gun-like Handheld Portal Device to place portals on exposed surfaces, allowing the player to traverse otherwise untraversable terrain. This dimension receives special attention within *LEGO Dimensions*, featuring prominently in the base game, constituting one of the five original Level Packs, and starring in the end credits.

Unlike most other media tie-ins in *Dimensions*, the *LEGO Portal* dimension plays with a franchise that has its own gameplay mechanics.[25] This encourages an unusual degree of self-reflexivity in which the story contradictorily positions the LEGO characters as falling into the *LEGO Portal* dimension while the gameplay subsumes *Portal*'s gameplay within the LEGO videogame genre. GLaDOS, the AI from *Portal*, operates within the diegetic world of the story, expressing surprise and indignation

[25] The other notable exception is the Midway Arcade level, which adapts the traditional LEGO gameplay into the style of various classic arcade games.

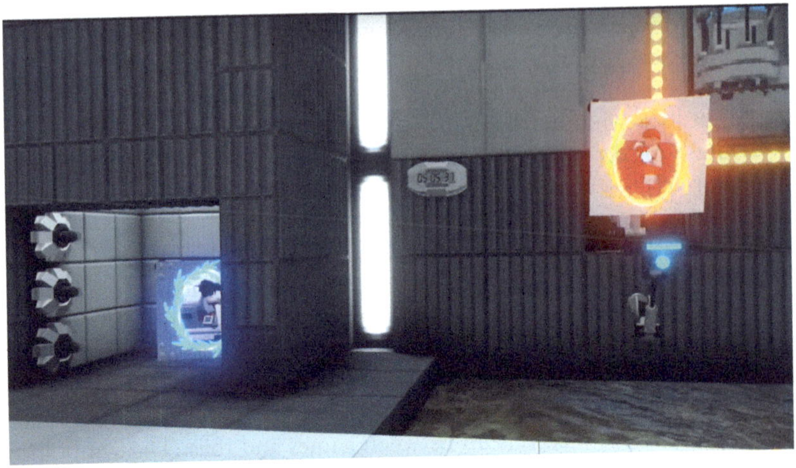

Fig. 4.4 Screenshot of minifigure Chell, the *Portal* protagonist, leaping through a portal in a trailer for *LEGO Dimensions*. While this gameplay and the industrial aesthetic of the level design resembles that of the original, key differences in the LEGO version include presenting the game in the third person and using portals as only one of many puzzle mechanics

when the protagonists solve purportedly impossible tests using abilities outside the physics of her reality (such as Gandalf's staff).[26] However, the levels are designed to be solved in precisely these ways as the underlying game logic remains entirely LEGO.[27] A game-within-a-game, the *Portal* dimension emphasizes the artificiality of the game construct to emphasize

[26] In one humorous instance, GLaDOS showers the LEGO minifigures with studs, saying "it seems you've been obsessively collecting these, so here's some more." GLaDOS is both aware of and able to manipulate these fundamentally LEGO elements without possessing any real understanding of them. Thus, whereas the LEGO characters believe they are trapped in another game but in fact are not, GLaDOS believes she is in her own game but in fact is not.

[27] Not only do the main characters not possess the portal gun around which the original games are designed, such a gun would not solve the kind of mechanical puzzles designed in the LEGO world. Even when the portal gun is unlocked through the Level Pack, it only latches onto special wall segments that appear at specific places within the level. In other instances, it reverts to a normal gun that can remotely 'smash' brick-built elements of the game world. It is still a valuable tool but is by no means the *novum* that it is in the *Portal* games.

play at or as the intersection of material and virtual, digital and analog, toy and game.[28]

Like *Portal*, LEGO adventure games present game progress as physical progress through a series of stages that can only be traversed if one 'solves' the spatial challenges of the level. Unlike traditional puzzles whose solution is an *answer*, the puzzles in *Portal* and *Dimensions* are resolved by *movements*: digital solutions must be implemented through analog gameplay, the constitutive comingling of digital puzzle and analog platforming that yields the hybrid genre of the puzzle/platformer. In general, to solve a puzzle is to discover a solution designed and embedded in the puzzle itself. Thus, to solve a puzzle is always to re-solve and thereby resolve it. Movement-based puzzles are somewhat different—rather than re-solving a puzzle within a purely informational space, movement-based puzzles require performing an activity that fits the scripted solution. This establishes a close parallelism between the narrative progression of the characters through the levels and the movement of the avatars through the levels. As solving a level puzzle unlocks the next level, thereby progressing the story, movement becomes both method and goal—the means to solving these spatial puzzles as well as the reward for solving them.

In *Dimensions*, the metaphor of the portal further adds narrative continuity to these level transitions. In many games, the loading screens that connect levels also fragment the represented space, revealing levels to be an artificial game construct rather than a direct representation of contiguous space. Like most LEGO adventure games, *Dimensions* similarly breaks immersion by using the loading screen between levels to evaluate the player performance in that level (mostly by measuring how much of various things players were able to collect or unlock). Narratively, however, the loading screens both into and out of different levels depict the inside of the portal as a kind of magical tunnel through which the characters fall. This makes the fragmentation of virtual space into disconnected yet connected levels more consistent with the story than just an

[28] *Portal* itself challenges the notion of the game as construct through gameplay which Daniel Johnson describes as calling into question decisions in which moving forward means displaying "obedience to the institution" (2009, n.p.) and a thematic presentation which Burden and Gouglas explain as "an algorithmic exploration of human struggle against algorithmic processes that have superseded their original intended purpose. The game explores the search for freedom from such computational processes" (2012, n.p.).

artifact of the puzzle/platformer genre. In addition, the portal animation provides a visual metaphor for the projection of the physical player into the virtual world (a similar effect is used in *The LEGO Movie* when Finn transports Emmet back into the LEGOverse). This analogizes how the game bridges material and virtual play by drawing a narrative of analog continuity that connects the digital fragmentation of the level structure.

This metaphorical projection also draws an analogy between player and player character, especially as *Dimensions* positions the player as simultaneously collaborating with, collecting, and controlling the minifigure characters. Similarly, the projective power of the portal allows the aforementioned trailer to dramatize the tension between narratively assisting but functionally controlling characters by showing McHale simultaneously supporting and exerting agency over[29] the minifigure characters. Built around puzzle mechanics that often require multiple unique characters' abilities to overcome, LEGO games are based on an ethos of teamwork that positions the player as coordinating interchangeable characters. From this position, the player plays *with* rather than *as* the characters. Like the dynamic of the character toy,[30] for which it is commonplace that character identity comes prepackaged such that players are encouraged to playfully reanimate formulaic narratives from the associated media, these games encourage the acting out of provided narratives rather than the creation of new ones. Similarly, the heavily scripted and linear narrative of *Dimensions* is framed according to a logic of toy play that positions players above rather than within the game fiction.

As a toys-to-life game that emphasizes its own mediateness, *Dimensions* presents virtuality as augmenting materiality and materiality as augmenting virtuality. Whereas material LEGO explores the digitality of the analog world by representing it through a fundamentally binary

[29] The trailer emphasizes the powerlessness of the minifigures due to difference in scale. McHale easily shrugs off even magical attacks, traps Batman under a mug, and picks up Wyldstyle, demonstrating his agency before using that agency to help rebuild the broken Batmobile.

[30] See Steven Kline's (1993) *Out of the Garden*, Dan Fleming's (1996) *Powerplay*, Marsha Kinder's (1993) *Playing with Power*, and Marc Steinberg's (2012) *Anime's Media Mix* for more examples.

system of atomistic connections, this game provides a hybrid material/virtual play that explores the integration of analog and digital experiences. Ironically, the playscripts that govern this experience engage digitality by analogy to LEGO toys without making much use of the digitality *of* LEGO toys. The various forms of digital bricolage—most notably, movement on the toy pad, puzzle-solving, and level progression—all add digital dynamics to analog gameplay independent of building with bricks. This digital analog, therefore, speaks not to the properties that remain constant as material LEGO is digitized into virtual LEGO, but to digitality as a space of interplay between multiple media which each reflect aspects of LEGO's brand identity.

Interplay

While these videogames both deconstruct the digital/analog binary, they do not simply invalidate it. Instead, they depend upon the *interplay* between these opposed yet compatible phenomenological paradigms. LEGO maintains its hybrid status as a digital analog not only by experimenting with crossovers between material and virtual worlds but also by cultivating a spirit of playfulness that seeks generative juxtapositions between digital and analog play. In this sense, it is better to refine the claim that LEGO *is* a digital analog by instead saying that the LEGO medium facilitates digital/analog (inter)play.

Such interplay, however, is not a pure, unmediated comingling of digital and analog. Despite their hybrid orientation, both games clearly differentiate digital and analog gameplay. In *Worlds*, the algorithmic nature of its digitized construction eliminates any analog building possibilities that the material toys have. This digitized construction takes place within the suspended frame of the tool interface, after which these constructions become fixed parts of the landscape upon which analog play unfolds. And in *Dimensions*, placing toys on the toy pad reduces their otherwise analog status to the on/off positionality of a switch. Then, the digital logics of puzzles and level progression organize analog gameplay. In both cases, virtual LEGO gameplay heightens the digitality already present in material LEGO toys, subsuming the analog characteristics of these toys within digital logics.

This gameplay is therefore in the spirit of the dominant ideology of LEGO toys in which a largely digital construction system designed to be mobilized for imaginative dramas. While construction play is not purely digital and dramatic play is not purely analog,[31] virtual LEGO positions dramatic play as less digital and more analog than for material toys. Although all virtual LEGO play relies on algorithms, only virtual construction play emphasizes its algorithmic nature. Instead, virtual dramatic play involves animated storytelling tied to the analog movements of player characters through an immersive three-dimensional virtual world. Thus, the symbolic bricolage that defines material dramatic play gives way to the world-building of rich virtual environments that reframe play from imaginative world-building into exploratory world-discovery.

By making construction play more digital and dramatic play more analog, virtual LEGO enhances the already extant sense that LEGO play is a two-stage process of first building and then playing within LEGO worlds. And although these examples of virtual LEGO are much less gendered than most LEGO play themes (*Worlds* because it is more thematically generic and *Dimensions* because it mixes such a wide variety of licensed themes), this separation of construction and dramatic play may nonetheless reinforce the gendered binarization of the medium.

However, it is certainly not the case that these games are less gendered because the virtual world somehow frees players from embodied or socially-constructed identities. To the contrary, virtual play is as embodied and socially constructed as material play, albeit in different ways. The socializing power of material LEGO, for instance, lies in embedded ideological playscripts that can suggest but not mandate particular expressions of play. The socializing power of virtual LEGO, by contrast, offers freedom of movement within the constraints of algorithmic worlds. In other words, to borrow the language of French theorist Gilles Deleuze (1992), material LEGO socializes by *discipline*[32] (ideologically conditioning individuals to self-regulate) and virtual LEGO socializes by *control* (algorithmically regulating individuals within surveillance networks).

Although these ideas characterize the changing power relations of the 'digital' age in ways that reflect the historicized and technologized use of digitality as computation, LEGO exerts these forms of influence in ways that resist a narrative of historical progression. While virtual LEGO is obviously more recent than material LEGO, it does not function as a

[31] Material LEGO toys have construction potential beyond the dominant Cartesian grid and dramatic play in LEGO involves the digital reassembly of symbolic bricolage.

[32] Deleuze is referring here to the theory of Michel Foucault, especially in *Discipline and Punish* (1975/1995).

new paradigm that reformulates the earlier relationship between control and freedom. Instead, within the LEGO transmedia empire, material and virtual LEGO reinforce each other—script each other's play. To characterize this emerging fusion of material and virtual LEGO play, it is worth restating the lesson learned by LEGO in their near financial collapse: that LEGO cannot thrive in a digital age by abandoning its core construction play, but rather by becoming a digital analog that hybridizes multiple forms of play.

In summary, LEGO's experiments in digital play are digital analogs not merely because they incorporate digital technology, but because they develop playful analogies between material toys and digital systems. Despite emerging from modernism, LEGO positions itself at the vanguard of postmodern toy design because its status as a medium of bricolage is particularly suited for developing digital/analog interplay. To perform bricolage is to bring discrete elements into new configurations, to create continuity from assemblages. Digital play in LEGO, therefore, refuses to align itself either with a material nostalgia in which tactile play reconnects the player with the fundamentality of body or with a transhumanist enthusiasm that the virtual will free players from the constraints of the body. Instead, postmodern LEGO play privileges bricolage as a hybrid experience of mediating material and virtual, analog and digital, object and system.

Embodied yet abstract, both material and virtual LEGO point toward and engage each other's reality—physical toys tangibly representing discrete design and computer-generated worlds smoothly moving and manipulating virtual objects. In so doing, the postmodern hybridity of LEGO bricolage brings together the dyads of play—construction and dramatic play, transmedia and attachment play—whose explorations flank this chapter. Moving beyond the colloquial distinction between digital and analog, LEGO consistently remixes elements of all its technological manifestations. The message of such multimodal play being, of course, the *medium* itself—that what LEGO means is bricolage.

Post-script 4—Digital Analogies in a LEGO Turing Machine

The phenomenological definition of digital and analog raises questions about the black-box nature of computer technology. While there are certainly important technical questions to be raised about how computers function, most everyday interactions with computers play out on their many surfaces—that is, using input and output devices classified as 'peripherals.' Consequently, modern users frequently experience a disconnect between their functional competence using computers and their abstract understanding (or lack thereof) of how computer technology works. In such an environment, cultural understandings of computers are often shaped by metaphors. While metaphors are most commonly conceived of as linguistic, computer interfaces are saturated with visual metaphors such as iconic representations of file folders, which both facilitate competence (that is, click on a folder to view its files) and imply how computers function (that is, hard drives operate like file folder). Yet, the everyday imaginary of what happens inside the black box of the computer can also be aided by material metaphors, which provide a uniquely *graspable* mode of conceptual representation.

Arguably LEGO's most educational toy, LEGO Mindstorms programmable robots allow learners/players to both physically piece together input and output devices in a mechanical system and provide a programming experience in which programmers can easily put together preconstructed blocks. Yet, whereas Mindstorms streamlines the programming experience in ways that reestablish the obscuring power of the black box, builders Jeroen van den Bos and Davy Landman of the CWI research institute used the Mindstorms NXT system to construct a LEGO Turing Machine (Fig. 4.5) that functions simultaneously as a working computational system and as a material metaphor for computation. Operating outside the dominant digital playscripts provided by LEGO, this build's innovative deployment of LEGO's digital analogs reveals the potentiality of theoretical/material interplay to tangibly manifest a thought experiment. In so doing, they show—not tell—how computers function by leveraging a toy designed for hands-on material play to model an otherwise black box computational system.

In 1936, Alan Turing published a mathematical foundation for a hypothetical computing device that later came to be known as a "Turing Machine." This hypothetical device computes by manipulating an infinite tape of squares, each able to contain a single symbol, according to a set of instructions for reading and rewriting the symbols and moving the tape left or right. Although its name suggests a tangible object, this device operates more as a thought experiment that still figures in modern theoretical computing, even though modern computers do not employ this specific technological model. Responding to this "purist" notion,

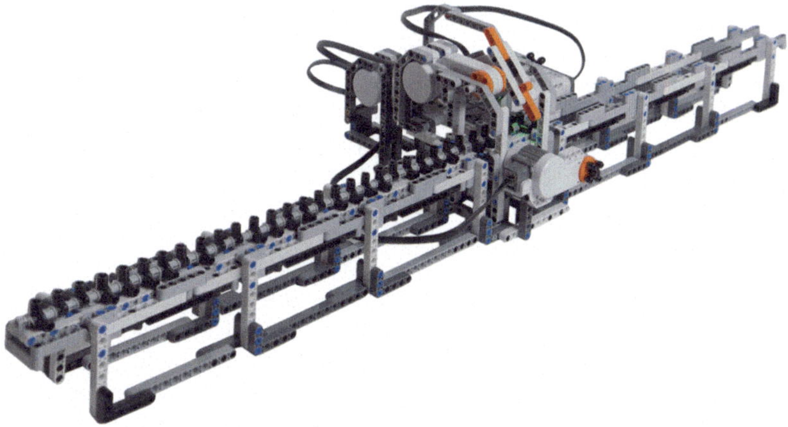

Fig. 4.5 *A Turing Machine built using LEGO* by Jeroen van den Bos and Davy Landman blogged by *Wired* in 2012 and posted on September 27, 2016 on legoturingmachine.org (now defunct) (Image courtesy of the artists)

philosophers Dresner and Rechter argue that "although Turing can be rightly described as drawing mathematics into the domain of mechanical symbolic manipulation, he should also be acknowledged as drawing symbols and symbolic manipulation away from the concrete or perceptible" (2016, p. 255). In other words, reducible to neither abstract mathematical theory or mechanistic physical form, the Turing Machine can be understood as establishing a complex interplay between theoretical purism and mechanical realism.

The LEGO Turing Machine takes this interplay one step further, adopting the material metaphor of the theoretical Turing Machine to present the informational aspect of computing in two ways that radically depart from the black box of the modern computer. First, it *mechanizes* the binary system of computer memory, literalizing the oft-used metaphor of the 'on/off switch' as representative of binary information. In this way, it resembles the LEGO System's materialization of atomism, which I describe elsewhere (2014) as literalizing the Democritean picture of atoms as tiny bits of matter connected with hooks and barbs, an image that continues to have metaphorical weight despite quantum theory showing that reality is far stranger. Second, the LEGO Turing Machine *externalizes* machine memory, inverting the typical black box and exposing its informational content to the world.

In both these ways, the LEGO Turing Machine is not designed for practical computing (after all, although the LEGO version is impressively capable of calculating 2 + 2, this 32-bit system has significantly reduced computing capacity compared with the Mindstorms brain that runs it). Instead, the LEGO Turing Machine deploys the mechanical properties of LEGO as digital analog to create a machine that becomes a material analog for digital computational theory. The primary point of this build, therefore, is that mechanical and exposed machine memory is more accessible to people trying to understand computing. As succinctly put by the builders: "abstract models are just that, an abstraction of something. In order to really show how simple the fundamental model of a computer is, we have developed a physical implementation of the Turing machine" (van den Bos and Landman 2012, n.p.). Thus, like the Turing Machine itself, this build blurs the line between idea and machine, redefining the computer as itself a fusion of material and virtual, digital and analog.

While any mechanical implementation of a Turing Machine might materialize this interplay, working with LEGO products heightens this digital/analog play. Using LEGO, moreover, allows the resonances of the LEGO brand to inflect this pedagogical moment. That LEGO is considered fun and easy helps demystify the intimidating abstraction of Turing's thought experiment. Furthermore, that LEGO Mindstorms cultivates an explicitly educational set of playscripts that encourages viewers to adopt its implied exploratory perspective.

Yet, of more interest here is how this build implicitly reconceives the scripts surrounding LEGO play. Despite its technological complexity, the Mindstorms system often presents a relatively simplistic picture of LEGO as digital analog. Mindstorms relies very little on the digitality of LEGO, reserving its core digitality to the programming of the computer and treating the mechanical LEGO Technic system (which relies heavily on connective systems based on rotation, making it much less Cartesian—and therefore less digital—than the LEGO System bricks) as the mechanical appendage to its robot brain. While the LEGO Turing Machine makes use of analog Technic components like gears and pins, it combines these components into a clearly digital system of binary switches. Thus, instead of aligning Mindstorms with robotics—an often-anthropomorphized field that, like virtual reality technology, uses digital computers to mimic analog movements—it uses the Mindstorms computer to reflect on its own computational nature. This draws an analogy between the otherwise separable Mindstorms "brick" (computer module) and Technic elements as potentially parallel digital or computational systems.

In so doing, this build also complicates the notion of a LEGO playscript by exposing the analogy between the performances of theatrical scripts and computational scripts. There is a clear parallelism between the use of the word *script* in the theater and in computing—the former directing an actor's performance by providing written dialogue and stage directions and the latter directing a computer to implement a series of instructions in a given order. Yet, there is a crucial difference in how these kinds of script organize these actions. Theatrical scripts are socially recognized as having an implicit imperative to be followed, but this imperative is not enforced by the script. Such scripts can be easily deviated from at the discretion (or error) of actors and directors. Computer scripts, by contrast, are implemented by computational systems incapable of reinterpreting the script in these ways.

While this distinction remains, the LEGO Turing Machine exposes how the black box of the Mindstorms brick implements its scripts. Every operation performed within the brick is given a physical manifestation in the movement of the switches and tape. Thus, although the Mindstorms brick is not capable of reinterpreting its instructions, the operation of the LEGO Turing Machine operates like a theatrical script in showing its viewers what executable operations look like, allowing individuals to imagine themselves in the role of the computer. After all, theatrical scripts and LEGO instruction manuals use the medium of print, a medium which is even more inexorably bound to its internal structure (the continuity of the print object) than computers (which perform operations temporally). As print communicates unchangeable scripts for changeable interpreters, this pedagogically-oriented material machine performs computing in ways that allow its viewers to play with and permutate the logic of computing.

LEGO—whether material or virtual—operates via playscripts that inevitably blend these two modes of scripting. Like computer scripts, LEGO instruction manuals provide unambiguous step-by-step instructions for achieving a desired result. Yet, unlike computer scripts these commands are not inherently compulsory. Like theatrical scripts, the many LEGO paratexts and representational content built into the LEGO elements themselves provides implicit imperatives about how play is designed to unfold but also ultimately leaves the particularities up to the player. In this mix of control and freedom, LEGO scripts establish themselves like ideological frameworks that do not compel but rather condition the significance of performed play. Yet, in the larger circulation of LEGO ideas beyond the core play supported by the LEGO company, builds like the LEGO Turing Machine provide new scripts for potential play that may be much more self-reflective than those that serve the LEGO brand.

Works Cited

Ackermann, Edith, David Gauntlett, Thomas Wolbers, and Cecilia Weckström. 2009a. *Defining systematic creativity*. Billund: LEGO Learning Institute.

Ackermann, Edith, David Gauntlett, and Cecilia Weckström. 2009b. *Defining systematic creativity in the digital realm*. Billund: LEGO Learning Institute.

Baichtal, John, and Joe Meno. 2011. *The cult of LEGO*. China: No Starch Press.

Bak, Meredith A. 2016. Building blocks of the imagination: Children, creativity, and the limits of *Disney Infinity*. *The Velvet Light Trap* 78: 53–64.

Bender, Jonathan. 2010. *LEGO: A love story*. Hoboken: Wiley.

Buerkle, Robert. 2014. Playset nostalgia: *LEGO Star Wars: The Video Game* and the transgenerational appeal of the LEGO video game franchise. In *LEGO studies*, ed. Mark J.P. Wolf, 118–152. New York: Routledge.

Burden, Michael and Sean Gouglas. 2012. The algorithmic experience: Portal as art. *Game Studies* 12 (2): n.p.

Coupland, Douglas. 1995. *Microserfs*. New York: Regan Books.

Cramer, Florian. 2019. What is "post-digital." In *New media, old media*, ed. Wendy Hui Kyong Chun and Anna Watkins Fisher, 689–702. New York: Routledge.

Deleuze, Gilles. 1992. Postscript on the societies of control. *October* 59: 3–7.

"digital, n. and adj." *OED Online*, Oxford University Press, March 2020, www.oed.com/view/Entry/52611. Accessed 16 April 2020.

Dresner, Eli, and Ofra Rechter. 2016. From symbol to 'symbol', to abstract symbol: Response to Copeland and Shagrir on Turing-machine realism versus Turning-machine purism. *Minds and Machines* 26: 253–257.

Fleming, Dan. 1996. *Powerplay: Toys as popular culture*. Manchester: Manchester University Press.

Foucault, Michel. 1995. *Discipline and punish: The birth of the prison*, trans. Alan Sheridan. New York: Vintage Books.

Goldberg, Daniel, and Linus Larsson. 2013. *Minecraft: The unlikely tale of Markus "Notch" Persson and the game that changed everything*, trans. Jennifer Hawkins. New York: Seven Stories Press.

Harel, Idit, and Seymour Papert (eds.). 1991. *Constructionism*. Norwood: Ablex Publishing.

Hills, Matt. 2016. LEGO Dimensions meets Doctor Who: Transbranding and new dimensions of transmedia storytelling? *Icono* 14: 8–29.

Johnson, Daniel. 2009. 'Lingua franca'—Portal and the deconstruction of the institution. Game Set Watch. http://www.gamesetwatch.com/2009/06/column_lingua_franca_portal_an.php. Accessed 16 April 2020.

Kinder, Marsha. 1993. *Playing with power in movies, television, and video games: From Muppet Babies to Teenage Mutant Ninja Turtles*. Berkeley: University of California Press.

Kline, Stephen. 1993. *Out of the garden: Toys, TV, and children's culture in the age of marketing*. London: Verso.

Lee, Jonathan Rey. 2014. The plastic art of LEGO: An essay into material culture. In *Design, mediation, and the posthuman*, ed. Dennis M. Weiss, Amy D. Propen, & Colby Emmerson Reid, 95–112. Lanham: Lexington Books.

Lee, Jonathan Rey. 2019. Master building and creative vision in *The LEGO Movie*. In *Cultural studies of LEGO*, ed. Rebecca C. Hains and Sharon R. Mazzarella, 149–173. Cham: Palgrave Macmillan.

A LEGO brickumentary [Film]. 2014. Directed by Kief Davidson and Daniel Junge, Radius.

LEGO Dimensions [Videogame]. 2015. Xbox 360.

The LEGO Group. 2014. A short presentation. LEGO.com. https://www.lego.com/r/aboutus/-/media/about%20us/media%20assets%20library/company%20profiles/the_lego_group_a%20short%20presentation_2014_english_ed2.pdf. Accessed 7 October 2014.

The LEGO Group. 2015. How to Play. LEGO.com. http://www.lego.com/en-us/dimensions/how-to-play. Accessed 25 February 2016.

LEGO Ideas Minecraft micro world—The forest [booklet]. 2012. Set 21102.

LEGO Worlds [Videogame]. 2015. Steam. 13 September 2018 Software Update.

Lévi-Strauss, Claude. 1966. *The savage mind*. Letchworth: Weidenfeld and Nicolson Ltd.

Mattes, Ari. 2019. Everything is awesome when you're part of a list: The flattening of distinction in post-ironic LEGO media. In *Cultural studies of LEGO*, ed. Rebecca C. Hains and Sharon R. Mazzarella, 73–95. Cham: Palgrave Macmillan.

Robertson, David C. 2013. *Brick by brick: How LEGO rewrote the rules of innovation and conquered the global toy industry*. New York: Crown Business.

Schut, Kevin. 2014. The virtualization of LEGO. In *LEGO studies*, ed. Mark J.P. Wolf, 227–240. New York: Routledge.

Steinberg, Marc. 2012. *Anime's media mix: Franchising toys and characters in Japan*. Minneapolis: University of Minnesota Press.

Stewart, Susan. 1984. *On longing: Narratives of the miniature, the gigantic, the souvenir, the collection*. Baltimore: Johns Hopkins University Press.

Thibault, Mattia. 2019. Transmediality and the brick: Differences and similarities between analog and digital Lego play. In *Intermedia games*, ed. Michael Fuchs and Jeff Thoss, 231–248. New York: Bloomsbury Academic.
Toups, Zachary O., Nicole K. Crenshaw, Rina R. Wehbe, Gustavo F. Tondello, and Lennart E. Nacke. 2016. "The collecting itself feels good": Towards collection interfaces for digital game objects. *CHI PLAY '16*: 276–290.
Turkle, Sherry. 1997. *Life on the screen: Identity in the age of the internet*. New York: Touchstone.
van den Bos, Jeroen, and Davy Landman. 2012. A Turing machine built using LEGO. Legoturingmachine.org. www.legoturingmachine.org/lego-turing-machine/. Accessed 27 September 2016.
Warner Bros. 2015a. LEGO dimensions—Alison Brie trailer. YouTube https://www.youtube.com/watch?v=efZjpArWNZs. Accessed 4 May 2020.
Warner Bros. 2015b. LEGO dimensions—Joel McHale trailer. YouTube https://www.youtube.com/watch?v=WXnOVyZaGK0. Accessed 4 May 2020.

CHAPTER 5

Story Toys: Transmedia Play in LEGO *Star Wars*

What makes toys magical—to me at least—is that their inanimate forms can conjure a profound sense of animateness, a seeming liveliness such that any remotely anthropomorphizable toy promises inexhaustible storytelling potential. Even extremely simple dolls can become beloved companions in countless imaginative adventures. And yet, on the surface, play and narrative media may seem incommensurable, the interactive, open-ended unfolding of the former precluding the linear, crafted textuality of the latter (and vice versa). This generates the constitutive paradox of the *story toy*: that toys which so clearly elicit storytelling are in some ways so incompatible with the licensed media narratives they adapt. As play and narrative are often intertwined in contemporary culture, this tension often boils down to a conflict between the contingency of playful narratives and the canonicity of licensed narratives. Consequently, *transmedia play*, in which licensed story toys allow players to imaginatively play out stories well beyond their canonical narratives, transforms the *dramatic play* discussed in Chapter 3. To deconstruct the ideology of narrative play most operative when the toy medium of LEGO performs its doubled role as a media or story toy, this chapter traces the tension between play and dis-play that runs through the media ecology of LEGO *Star Wars* toys from its playsets to its transmedia paratexts (specifically television serials and videogames).

LEGO has exploded from a singular toy medium into a vast transmedia empire with successful incursions into most major popular media

J. R. Lee, *Deconstructing LEGO*,
https://doi.org/10.1007/978-3-030-53665-7_5

forms, including books, music, film, television, board games, videogames, magazines, apps, merchandise, and more. Not merely investing in these individual properties, LEGO exemplifies transmediality by developing a strong branded synergy across these disparate media forms. In a sense, LEGO is perfectly poised to mediate a transmedia phenomenon, possessing the unifying potential of an iconic, ideologically-coherent brand alongside the remixability of a medium of symbolic bricolage. After all, *transmedia* primarily names a corporate storytelling practice that develops synergy across multiple media, and LEGO is a rare blend of brand *and* medium. LEGO, that is, both contributes iconic branded content to transmedia phenomena and contributes a site for transmedia phenomena to play out.

The multifaceted characteristics of this toy medium and media toys have therefore allowed LEGO to develop several successful original transmedia worlds—Friends, Ninjago, Bionicle, etc.—as well as dozens of successful tie-ins with other media franchises, including *Harry Potter*, both *Marvel* and *DC* comics, *Lord of the Rings* (Fig. 5.1), and *Star Wars*. The most extensive of these partnerships is LEGO *Star Wars*, which launched in 1999, instigating what Lars Konzack identifies as the era of transmedia LEGO (2014, p. 8). Still thriving more than two decades later, LEGO *Star Wars* currently extends across hundreds of sets and almost[1] all the aforementioned media. At its twentieth anniversary, LEGO *Star Wars* had reached "Around 700 different sets (including 1000 LEGO *Star Wars* minifigures), five video games, several television and web series, and an abundance of other fun merchandise" (LEGO 2019, n.p.). LEGO has managed to embrace co-branding despite—and even because of—the strength of its own brand identity with its ethos of unifying mixed messages.

While in retrospect the expansion of LEGO into a transmedia phenomenon seems inevitable, LEGO was surprisingly slow to embrace its transmedial potential. Even its monumentally successful partnership with *Star Wars* was somewhat belated, coming two decades after LEGO released its first modern minifigures and play themes in 1978. This same

[1] LEGO *Star Wars* does have presence in all the aforementioned media, but has more minimal presence in music and film, as one could exclude the soundtrack for the videogames as being not particularly connected to LEGO (unlike the songs and musical numbers in the two LEGO movies) and the short *Star Wars* cameo in *The LEGO Movie* as being too incidental.

Fig. 5.1 Full-page advertisement from the LEGO Club Magazine advertising the LEGO *Lord of the Rings* videogame. This self-reflexive advertisement shows the LEGO brand as a unifying force for creating transmedia synergy. Visually representing the LEGO ethos of connectivity, this advertisement shows the videogame uniting mortal enemies in friendly, self-reflexive gameplay. Furthering the self-reflexivity, a copy of the LEGO Club Magazine appears on the coffee table. As the advertisement depicts the magazine it is contained in and the characters play a videogame they star in, LEGO playfully advertises a self-aware transmedia experience. Here, the characteristic humor of LEGO leverages its cartoonishness—not to promote immersion—but to promote playful, nostalgic extensions of the transmedia world. From a corporate perspective, this ad also shows the co-branding behind transmedia storytelling by juxtaposing the LEGO and *Lord of the Rings* logos at the top and arraying the five production company logos at the bottom

year, Kenner released a line of collectible licensed *Star Wars* toys in a moment which toy scholar Gary Cross describes as one of the critical events in the history of transmedia play (1997, p. 190). And LEGO lagged behind the next wave of transmedia toys as well, only fully embracing its status as a licensed story toy well after the craze for toy-based 1980's television series like *Transformers*, *G.I. Joe*, *My Little Pony*, and *Jem and the Holograms* had long since faded away. Instead, during this time LEGO experimented with story toys in ways that "bucked the trend of action-figure fantasy" (Cross 1997, p. 220). Yet, while LEGO is late to the history of story toys, it does offer a rather distinctive take on transmedia play worth exploring as a significant complement to this history.

As a toy medium and media toy, LEGO does everything story toys traditionally do to extend transmedia franchises while also pioneering a uniquely LEGO-like form of storytelling. As Derek Johnson argues, LEGO became "One of the biggest success stories for co-franchising" (2013, p. 234) by flaunting its own brand identity even at the expense of continuity with its source text. In the case of the LEGO videogames, Johnson notes that "Though Lego was the licensee for each of these Hollywood properties, the company used its limited claim to those resources to build a separate video game franchise defined as much by the Lego name and aesthetic as by the media license" (p. 234). Similarly, Neal Baker writes that, beyond mimesis, licensed LEGO characters "represent the film source, certainly, but via stylized, LEGO conventions that adhere to a recognizable, global corporate brand more than they imitate the live-action figures of the film" (2014, p. 43). In other words, LEGO has constructed an entire paradigm of transmedia storytelling in which narrative and play are intertwined according to the pixelated playful bricolage that defines the LEGO medium.

Yet, LEGO does not only adapt transmedia narratives into distinctly LEGO versions: it also transforms itself into a distinctly transmedial medium. This is something no other medium does quite like LEGO. In hundreds of sets, LEGO *Star Wars* has introduced so many distinctly *Star Wars* elements that LEGO has now become a medium materially inflected with *Star Wars*. As a medium that has become saturated with deconstructed bits of storyworlds, LEGO materializes the symbolic elements of storyworlds to itself become a transmedia world.

And, although this is certainly more prevalent within its licensed product lines, the interchangeability of LEGO elements means that these fragments of story material circulate throughout the medium, sometimes appearing in somewhat surprising places. According to the bricklink database, the LEGO Lightsaber Hilt (Item #64567) that represents the iconic Jedi weapon introduced in *Star Wars* appears in 658 sets, almost two-thirds of which are *not* associated with *Star Wars*. This includes no less than fifteen LEGO Friends sets and eight Modular Building sets (including the Parisian Restaurant described in Chapter 1).[2] If, as this project argues, LEGO is a medium of bricolage designed for the creative reassembly of *already-significant* elements, then LEGO is also a medium of transmedia bricolage whose already-significant elements are scavenged from its licensed tie-ins.[3] Consequently, transmedia LEGO play represents a new paradigm for participating in transmedia world-building that speaks to the growing entanglement of media and play culture.

After theorizing the tension between play and dis-play in the context of the fraught position story toys have occupied within contemporary media studies, this chapter traces how this dynamic plays out in the parallel realms of the toy medium and media toy. First, as a toy medium, LEGO blends construction and dramatic play to construct storytelling and world-building as bricolage, reassembling symbolic elements drawn from licensed narratives into story toy play. Second, as a media toy, LEGO surrounds its construction toys with some unusually substantial media texts that do not merely extend *Star Wars* but instead create a distinct LEGO *Star Wars* world that is related to but nonidentical with the canonical *Star Wars* universe.

This exploration ultimately reveals a strange irony symptomatic of the constitutive paradox of the story toy—that the playscripts embedded in the mobile, reconfigurable LEGO toys typically reinforce the canonical narrative while the playscripts embedded in the more narrative media are typically more playful. That is to say, the LEGO *Star Wars* toys present themselves as more film-like than they actually are whilst their other media texts present themselves as more toy-like than they actually are.

[2] Based on a bricklink.com database search conducted March 2020. The number of *Star Wars* sets is 248 (or 37.7%). The count of LEGO Friends sets combines the tags "Friends," "Disney Princess," "DC Super Hero Girls," and "Elves."

[3] In an analysis of two original LEGO transmedia franchises—Ninjago and Chima—Lori Landay analyzes storytelling based on "myth blocks," a version of symbolic bricolage that operates "as if someone had broken apart mythologies into parts and they had landed in the toy box, myth blocks there to be configured however one wished" (2014, p. 55).

This suggests that although transmedia storytelling is a media-specific endeavor, it may sometimes proceed from an implicit denial of the medium as if suppressing the specific character of a medium's preferred mode of storytelling allowed it to achieve a narrative continuity it might otherwise struggle to actualize. Rather than precluding transmedia play, however, these parallel negations are synthesized within a constitutive tension between play and dis-play that produces the narrative potential of transmedia toys.

Transmedia Play and Dis-Play

Although story toys are some of the largest players in transmedia worlds when measured by investments of time or money, transmedia play is very intermittently and inconsistently analyzed within the most influential transmedia theories. To contextualize this analysis, this section takes a brief foray through a few significant trends in transmedia theory before proposing the constitutive tension between play and dis-play as a way of reimagining what story toys contribute to transmedia worlds.

To make a somewhat sweeping generalization, it seems that the greater emphasis a transmedia theory places on storytelling, the less attention it tends to direct toward story toys, which are not considered as offering sufficient narrative content to do more than supplement more narrative media. For instance, in her influential *A Theory of Adaptation*, Linda Hutcheon argues that no clear dividing line distinguishes adaptations from non-adaptations while—without further theorization—placing franchised toys like Barbie's Titania and Galadriel on "this 'expansions' end of the continuum" (2006/2013, pp. 171–172). Similarly, in his equally influential *Convergence Culture*, Henry Jenkins (2006) considers toys not to be full participants in *transmedia storytelling*, as he names the movement in contemporary media culture toward the increasing convergence of media franchises.

These exclusions make perfect sense if one considers that these theories aim to explain how narrative texts circulate within public media cultures that cannot easily capture the fleeting, private performances of narrative play. Toy play does not produce circulatable texts along the lines of most other fan media (such as the fan videos that Jenkins discusses). Nevertheless, toys certainly lend their distinctive textuality as scriptive things to media fandom. As playable consumer goods, story toys embody key aspects of the kind of corporate media culture that theories of adaptation and transmedia storytelling aim to analyze. Including toys in discussions of adaptation or transmedia storytelling need not claim that toys mediate storytelling in exactly the same ways as other more traditional media. Including toys simply acknowledges that toys do in fact participate in the broader cultural formation of transmedia phenomena, both for corporate entities who strategize narrative synergy across toys and other media and for consumers who link their toy collection and play to their consumption of other media. LEGO is a medium of transmedia play, therefore, because one defining orientation of its play is participation in this broader transmedia culture.

Toys figure much more prominently in theories that privilege material culture over storytelling, such as work by Steven Kline (1993), Ellen Seiter (1993), and Dan Fleming (1996) on media toys. More recent theories that apply some of these insights more directly to transmedia phenomena include Lincoln Geraghty's (2014) *Cult Collectors*, Jonathan Gray's (2010) *Show Sold Separately*, and Marc Steinberg's (2012) *Anime's Media Mix*. Geraghty documents how cult collection functions as a unique form of media convergence in which consumption of toy objects serves the production of fan identities. Gray argues that toy *paratexts* "play a constitutive role in the production, development, and expansion of the text" (2010, p. 175), a claim made possible by his understanding "text" as an experience/performance that transcends the recorded story. And Steinberg traces how toys figure more prominently in Japanese *media mix*, which is "roughly equivalent to the North American term, transmedia storytelling, with the significant caveat that the media mix is not tied to stories, as it is often developed around characters" (2012, p. 144). Together, these theories offer valuable paradigms for studying media as more than a vehicle for storytelling, but instead as a means of *making* or constructing things—be they fan identities, texts which are "only experienced in the act of consumption" (Gray 2010, p. 30), or entire cultural experiences of media mix.

Yet a further link is needed to reconcile storytelling and making within story toys. Theories of *transmedia world-building* offer an avenue for doing so. Klastrup and Tosca define "transmedial worlds" as "abstract content systems from which a repertoire of fictional stories and characters can be actualized or derived across a variety of media forms" (2004, p. 409). Less starkly, Mark J. P. Wolf writes that "completeness of a world is what makes it seem as though it extends far beyond the story, hinting at infrastructures, ecological systems, and societies and cultures whose existence is implied but not directly described or clearly shown" (2012, p. 42). Transmedia worlds are conceptually abstract realms that often imply storytelling while being themselves non-narrative. Thus, in a certain sense, story toys are like the material manifestations of transmedia worlds, tangible objects that similarly imply storytelling while being themselves non-narrative.

Indeed, such theories do not so much bridge the gap between storytelling and making, but rather reconstrue storytelling as a form of making. Thus, for Wolf, building transmedia storyworlds is an act of *subcreation*, a term Wolf borrows from J. R. R. Tolkien to describe how human storytelling imperfectly reflects the world-shaping of the Creator God. In the terms of this project, subcreation can be conceived as an act of narrative bricolage that creatively reassembles already-significant elements drawn from an already created world—not only the world of int(e)ractable sensory phenomena but also the ideologically, culturally, and narratively inflected worlds through which such phenomena are mediated. Consequently, what is at play in transmedia play is how transmedia worlds are materially instantiated within tangible, interactive media that allow players to simultaneously enter into subcreated fictional worlds and participate in their ongoing creation.

Story toys, then, play a special role in materially mediating conceptually abstract storyworlds, without dictating particular stories. Much more open-ended than even the most sandbox-style videogame, story toys are spatially rather than temporally oriented texts that stage possibilities for storytelling. As material possibility spaces, story toys concretize imaginative bricolage by tying storytelling potential to the mobile signifiers of the material toy system. In other words, the very non-narrative qualities that situate story toys outside transmedia storytelling anchor their potential to materialize transmedia world-building. As storyworld philosopher Marie-Laure Ryan puts it, "non-media objects such as T-shirts, Lego sets, and costumes that can be used by fans as 'props in games of make-believe'" (2017, p. 38). This notion of *props* as fixed yet mobile signifiers for imaginative play references the work of Kendall Walton (1990), whose titular theory of *Mimesis as Make-Believe* positions all representational arts as material (i.e. mediated) props for the experiential and imaginative construction of stories through *games of make-believe*. This suggests that media consumption is always an act of production insofar as the narrative imagination plays out stories with and within the mediating force of media texts. Under this paradigm, therefore, story toys materially manifest the theoretical bricolage that underlies the representational arts, functioning as literal props that actualize the primary metaphor for the narrative imagination.

Materially actualizing a storyworld, however, simultaneously entails media-specific gains and losses. The material characteristics of any prop necessarily constrain the play potential it affords. This is true of any media—not only story toys. What makes story toys special is that their materiality is of a kind with the physical worlds they represent. Most media[4] privilege the more perceptual senses of sight and hearing over the more bodily, proprioceptive, and even intimate sensations of touch. Consequently, story toys are not just material cues to imagine conceptually abstract storyworlds—story toys offer fully functioning physical worlds unto themselves.[5] Materialized in story toys, fictional worlds become fully realized as genuine albeit miniaturized worlds.

At the same time, this realization complicates the fictionality of such worlds. Whereas some media function more like windows into unreachable worlds (and often ask participants to forget their materiality as they immerse in the narrative experience), toys require a parallelism between fully real playspaces and the fictional worlds that are imaginatively layered onto these contiguous spaces. Toys demand that these two layers of significance be collapsed into each other, as when playing house nests a fictive domestic space within an actual domestic space (see Chapter 2). Yet, toys also require that these layers remain at least somewhat separate, as the

[4]The most notable exceptions being architectural experiences like escape rooms and theme parks, and their virtual counterparts of videogames and virtual reality. These forms could be imagined as spatial inversions of the story toy as they largely involve immersing the player within a physical environment, whereas toys typically place the player outside and/or above the manipulable miniature playspace.

[5]It might sound odd to consider a doll or action figure a world unto itself when people generally talk of bodies as agents that move around within larger worlds. And yet, the distinction is not so strong as it seems. As Kant (1770/1929) argued, there is a sense in which spatiality is extrapolated from the body. The world does not come pre-inscribed with a Cartesian coordinate grid that measures its spatial extensity. Instead, the human capacity for bodily movement orients our understanding of the world—the possibility of navigating a space, a potentiality that requires the body, is what defines our spatial understanding. Similarly, I understand the spatial extensity of the doll or action figure as linking the physical world of imagined play with the material world of the player in a way that, for instance, the impassible fourth wall of the film screen does not.

player cannot fully immerse in the fictive realm but must instead maintain some awareness of what lies outside the *magic circle*[6] to continue performing toy play.

Transmedia play adds a third layer to this underlying duality: awareness of the represented source narrative. Story toys require an unusual degree of awareness of their circulation as commodities since their status as licensed collectibles is part of their narrative potential. This is particularly true for LEGO, at once a toy, medium, and brand. And yet, story toys promise to transcend their commodity status as mobile props for games of imaginative make-believe. That is, story toys are constituted by a tension between *play*—the activation of the toy in dynamic, playful performance—and *dis-play*—the pacification of the toy into a static representational commodity. As various aspects of the potentiality of play are already explored in the previous chapters, this section primarily explores how dis-play constitutes a shadow complement to play, grounding its animateness in the inanimate.

To *display* an object is simply to stage it for visual consumption. Consequently, in its strongest sense, to *dis-play* an object is to remove it from circulation in play to instead circulate as a meaningful participant in a collection, a curated assemblage of objects that perform a predominantly visual rhetoric. Objects on dis-play are transformed from materially circulating within culture into fixed visual signifiers *of* that culture (as when artifacts in a museum display signify long-lost or faraway cultures). As Susan Stewart puts it, "the collection represents the total aestheticization of use value" (1984, p. 151), transforming functionality from interactive potential to visual performance. In an inherently nostalgic fashion, such objects perpetually point to a bygone past whose unreachability is symbolized by the material separation that defines display as dis-play. As licensed collectibles,[7] transmedia LEGO toys can certainly promote dis-play in this

[6] This term refers to how play almost magically circumscribes a space that suspends certain features of everyday life to instead create a new reality that operates according to the rules of play. For instance, realistic interpretations of the environment may give way to fictive ones or certain kinds of possible action are suspended to conform with game rules. This term derives from the work of Dutch historian Johan Huizinga (1938/1950) in *Homo Ludens* and has become a useful albeit limited metaphor.

[7] Dis-play has been a key feature of toydom at least since the rise of modernity, in which to be a toy became also to be a collectible. While the cultural practice of collecting toys in their original packaging is more recent, toy history is saturated with the specter of collection. For instance, it is difficult to historically disentangle the parallel rise of

most extreme sense. Beyond this, however, all transmedia play in LEGO is characterized by an emergent tension between play and dis-play as a logic of collectability inflects playful storytelling with the spirit of conspicuous consumption. After all, most toys circulate through collections in which they are at times in play and at times on display, the specter of one mode haunted by the other.

Not only do toys often move between play and dis-play, they perform both modes as shadow complements of each other. Playful performances are often staged according to a visual logic of display—after all, the theatrical design of playsets discussed in Chapter 3 depends on implied spectatorship. Conversely, collections on display are often staged in ways that evoke the dynamic, animating movement of play—for instance, the static photographs on LEGO packages are almost universally posed as *in medias res* action shots. Play and dis-play are intertwined performances akin to the intertwining of interact-ability and intractability that constitute int(e)ractability. More than a medium of mixed messages, LEGO mixes play and dis-play as co-constitutive forms of mediateness.

In this spirit, I argue that transmedia play with story toys both creatively reassembles dis-played signifiers and itself constitutes a signifying performance oriented toward display. That is, transmedia play draws up a reservoir of established signifiers to lend their authenticity and materiality to imaginative play while deploying these signifiers in an embodied performance that implies the possibility of spectatorship. Thus, while the presence of dis-play always threatens to arrest and negate the playful performance, it paradoxically also animates play by materially linking it with the transmedia storyworld. Like many dynamics at play in LEGO,

dollhouses as children's playthings and miniature houses as spaces for adult collection and display. And it is similarly difficult to disentangle the marketing logic of collectability from the design of toys organized into product lines or given rarity through limited editions or cereal box mail-ins. In fact, this is precisely the thinking that gave rise to the development of LEGO itself, as recounted in the origin story of how Godtfred Kirk Christiansen became inspired to develop LEGO after a conversation with a retailer who lamented that toys were difficult to sell without an underlying system (see Preface). As LEGO has evolved, the system has been increasingly designed and marketed to be collectible. As Lincoln Geraghty (2014) describes, LEGO uses a variety of strategies to cultivate collection, including publishing checklists and guides for collectors, releasing minifigures with differing levels of rarity, and offering massive display models in the *Star Wars* Ultimate Collector's Edition sub-line.

this too expresses a fundamental int(e)ractability as the intractable signifiers that fix the relationship between the toy and the storyworld become mobilized within interactive storytelling play.

Dis-Play in Play: LEGO Story Toys

As objects that signify the fictional worlds underlying transmedia storytelling, story toys put dis-play into play by first actualizing and then mobilizing fictional worlds. Story toys actualize fictional worlds by translating conceptual abstractions into a material and therefore commodifiable objects. Story toys lend material reality to fictional worlds as spatially rather than temporally oriented media. As Madeleine Hunter notes,

> Lego's franchise themed playsets frequently enact a transposition from a temporal medium to a spatial medium. Rather than adapting the narrative per se, these playsets aim to provide players with the pieces necessary to construct representations of the characters, settings, and props associated with specific narrative sequences that the player is then invited to re-enact. (2018, p. 277)

Story toys are not narrative performances themselves—instead, they are material and spatial props for creating temporal narrative performances. These temporal performances then mobilize these props, reassembling their already-significant elements into a narrative bricolage.

Consequently, the main subcategories of story toys—character toys, vehicles, and playsets—lend their material reality to representations of the characters, technologies, and settings that anchor fictional worlds. Story toys then draw upon their status as toys—objects to be toyed/played with—to mobilize these object-centered material realities, encouraging consumers to become players by circulating these representational objects within narrative play. This stage of playful enactment is, notably, contingent. A story toy perpetually on display might never see actual play. Yet story toys imply play even when on display—their embedded scripts are *play*scripts that signify narrative potential. This section, therefore, deconstructs the playscripts woven through LEGO *Star Wars* character toys, vehicles, and playsets to trace how these designs render fictional worlds int(e)ractable by both offering the signifying elements for bricolage and invoking the logic of the source narrative as a structural principle for narrative play.

Character Toys

As *character toys*—toys that embody playable characters or personalities—LEGO minifigures spatialize characterization by encoding identity into their material designs. These toys object-ify characters, transforming them into crystallized narrative archetypes. In general, character toys do not represent autonomous agents the way characters in other media often do, enacting their own choices as the audience spectates. Instead, somewhat like first-person narrators in literature, character toys waver between being representations of fictional characters and functioning as open subject positions for players to insert themselves in the storyworld.

In this vein, Dan Fleming argues that the Kenner *Star Wars* action figures that helped keep the franchise relevant in the gap between the first and second film release were successful in part because they featured a relatively blank protagonist:

> Luke Skywalker, the young hero of the film trilogy, is actually a rather softly defined character. His filmic identity and self-realisation are defined by the two contexts in which he operates: the loose coalition of misfits, adventurers, political idealists, robots and aliens marshalled against the evil Empire; and the ever-present technologies on which survival depends in that hostile galaxy. (The dense elaboration of such contexts can be called 'metonymic' in rhetorical terms.) The character's interactions within these two contexts give substance to his identity, rather than any more conventionally individualistic representation of character traits or private compulsions. Identifying with Luke, especially for a child, will consequently depend on imagining oneself in those contexts rather than isolating Luke or actor Mark Hamill from them as an object of reified identification. (1996, p. 96)

In this moment of identification, the character functions as the avatar through which the player can immerse in the fictional world. This immersion rests on a delicate balance between blankness and specificity, with blankness helping players more easily identify with the avatar, and specificity helping explicitly link the avatar to its narrative context.

Given their role as avatars, the representational strategy of character toys is to translate subjects into *subject positions*—that is, into playable roles. Character toys break characters out of fixed roles within canonical narratives to redeploy them as mobile signifiers for emergent narrative play. In other words, character toys transplant characters from stories into story*worlds*, from narratives into narrative possibility spaces. As Marc Steinberg notes of the character toys that anchor Japanese media mix,

> The manga and anime series gave the toy a personality, a narrative setting, a group of characters, a series of set poses, and a voice. In short, the manga and anime character gave the toy a "world." They also broadcast this character-world on a mass scale. In return for its accession to a world, the toy gave the character matter, narrative openness, and movement. (2012, p. 122)

Lending a character "matter, narrative openness, and movement" transforms how character functions. The mobile commodity of the character toy is a signifying object that scripts narrative play according to an archetypal logic. Character toys typically depict not a character's canonical actions but rather the character's disposition to act in certain ways. Luke Skywalker action figures, for example, are less likely to depict full narrative scenes and more likely to depict Luke as the embodiment of a particular narrative function, more precisely the heroic Jedi Knight who bears a responsibility to restore moral order to a chaotic universe by drawing on the mystical energy of the Force. In this way, such toys deemphasize Luke's personality and instead emphasize Luke's positionality as an archetypal and playable subject position with a clearly defined narrative role.

This is a radical inversion of the narrative logic of most traditional storytelling media, which typically narrate a single, fixed progression of events. In such narratives, fictional characters typically function as agents that shape and subjectivities that interpret these unfolding plots. This function identifies characters as subjects that transcend their contingent[8] plotlines analogous to how the identities of real people exceed their individual circumstances. By contrast, character toys rarely posit internal identities that exceed characters' contingent roles within canonical narratives. Instead, while unhinged from any single fixed narrative, the media-specific way character toys encode identity elevates contingent, external markers like outfits and accessories to essential characteristics. As

[8] Literary theorist Andrew Miller (2008) argues certain kinds of fiction heavily rely on an *optative mode* that weighs the (usually ethical) consequences of paths taken with the potential consequences of alternate paths not taken. By lending material reality to characters beyond their integration into plotlines, character toys make it possible to playfully pursue paths already implied in the optative mode of canonical narratives.

character toys have no consistent way of making characters identifiable without directly signifying or displaying such trappings, they are often more *iconic* than representational.

When theorizing comics, Scott McCloud (1994) proposes the *icon* as a way of understanding how images in comics can function somewhat like language, signifying based on recognizable associations rather than realism. Similarly, Jessica Aldred notes of the minifigure aesthetic in LEGO videogames, "the most basic iconography of character becomes key to player recognition, boiled down to one or two essential traits that can be ported across media" (2014, p. 110). Through this iconic representation, otherwise contingent, external characteristics become elevated into essential identity conditions, constituting rather than merely characterizing the character. Consequently, whereas other narrative modes can portray Han as a smuggler who might have been otherwise had circumstances been different, character toys make signifiers of smuggling immutable, material conditions of being Han. Without fixed narratives, toy versions of Han may paradoxically escape ever participating in smuggling even while they cannot escape being identified by embedded scripts that signify smuggling. Character toys thereby draw upon canonical imagery in ways that elevate these images to iconic status.

Since they are not narratives themselves, toys are seldom considered canon. Yet toys play an influential role in canonizing the narratives they represent. Character toys assemble characters from a symbolic bricolage of recognizable, displayed elements drawn from canonical narratives. In so doing, they reinforce the very canonicity they rely upon. Although all officially authorized narratives are to some extent canon, narratives that are more consistently referenced as canon typically achieve heightened cultural status. Since their representational strategy is based on the icon, character toys play a special role in canonizing the visual icons that define iconic characters. To borrow Aristotelian language, this transforms small visual details from *accidental* to *essential* properties—that is, from properties that a thing just happens to have to properties that make something what it is. Toys elevate contingent design details like Han's vest, Leia's hair buns, and Luke's lightsaber into identifying features.[9]

[9] To this day, the most canonical image of Han (as evidenced by how he is represented in various paratexts including fan art and cosplay) is a black vest over a gaping white shirt, dark pants with a stripe running along the outside seam, and a brown leather belt and blaster holster. While this character design is present in the films, its canonicity is

While all character toys can canonize visual icons in this way, LEGO minifigures have a distinctive aesthetic of cartoonish abstraction that may enhance the function of the icon. McCloud describes cartooning as "a form of amplification through simplification" (1994, p. 30), a description that easily applies to LEGO. In a similar vein, Mark Wolf argues that the locations in the LEGO *Star Wars* Death Star set "are caricatures of the films' locations, simplifying them and exaggerating their salient features" (2014, p. 27) and Aldred argues that "LEGO video games mobilize what Donald Crafton calls a 'figurative' mode of cartoon performance, characterized by its extroverted style, formulaic character types, and recurrent gestures, sayings, and gags" (2014, p. 114). Materializing fictional worlds not as perfect replicas but as abstractions and exaggerations, character toys signify the *idea* of the character in ways that construct archetypal versions of the characters they refer to.

Nowhere is this archetypical abstraction more clearly illustrated than in the iconic LEGO minifigure face. Not only are these faces more cartoonish than typical for action figures, but they are also literally pictorial cartoons insofar as they are simple yet evocative two-dimensional images printed upon the surfaces of cylindrical minifigure heads (see Chapter 3). These faces replace the nuanced expressions of film actors with simplified expressions that may better reflect the affective intensity of fan response.

The more recent trend toward minifigure faces that reflect clear emotions employs exaggerated icons that signify pure, intensified versions of emotions, creating a well-defined playscript for narrative play. And the older, more established trend of relatively blank minifigure faces invites players to identify with generic subject positions, as when McCloud writes that "The cartoon is a *vacuum* into which our *identity* and *awareness* are *pulled* … an *empty shell* that we inhabit which *enables* us to travel in *another realm*. We don't just *observe* the cartoon, we *become* it!" (1994, p. 36) Between these two aesthetics, slightly personalized minifigure faces served as icons that simultaneously identified and characterized their characters. For example, the sideways smirks and wry smiles of roguish smuggler Han Solo's early minifigure designs portrayed Han as

solidified by the Kenner *Star Wars* action figures, which made it easy to pause and reflect on visual details it is easy to overlook within the dynamic action of the film. Thus, the red stripe went from an incidental detail in the films to an icon in the toys to a named badge of honor—the Corellian Bloodstripe—in the later Expanded Universe texts.

an archetypal rogue/cynic, amplifying his archetypal filmic persona. Once again, blankness and specificity are intertwined within a signifying system designed to facilitate the similarly intertwined practices of transmedia play and dis-play.

The most notable exception to LEGO's cartoonish abstraction is the replacement of the ubiquitous yellow skin tone with a range of more realistic skin tones. Such specificity plays into the complex politics of the aforementioned affective-character of the represented storyworld. Unlike the way "ethnic" Barbie dolls commodified generalized racial stereotypes,[10] brown skin in LEGO is inextricably tied to licensing agreements that necessitated minifigure versions of non-white bodies—initially NBA athletes, and shortly thereafter Lando Calrissian from *Star Wars*. As others have noted, moving toward 'realistic' skin tones to represent Black personalities clearly demonstrates that the hitherto ubiquitous yellow skin has always implicitly signified whiteness.[11] More precisely, yellow has always meant 'race-neutral' within a cultural ideology of 'neutrality' based on the invisibility of whiteness as the norm and the visibility of all other forms of 'color' as marked difference. This complicates the availability of different characters for role-play by racializing blankness itself. In other words, part of the 'blankness' that Fleming attributes to Luke is the assumption of an unmarked whiteness that makes this farmboy from outer space feel culturally 'relatable' to a presumed white American audience. Consequently, how certain identities are differentially dis-played also informs how available such characters are for narrative play.

All these ways that LEGO character toys materially manifest the tension between blankness and specificity contribute to embedded playscripts for transmedia play. The fixed significations placed on dis-play are designed to

[10] Ann duCille traces the complex relationship between representation and play, noting for instance that ethic Barbie dolls "are designed and marketed at least as much with adult collectors in mind as with little girls" (1996, p. 42) and that "If black children continue to identify with white images, it may be because even the would-be positive black images around them—including black Barbie dolls—serve to reinforce their second-class citizenship" (pp. 47–48).

[11] John Baichtal and Joe Meno record that among the fan community, "There was also a general perception that the yellow color of the minifigs actually signified a Caucasian" (2011, p. 59). Derek Johnson notes that "If yellow could stand in for the white norm from which Lando was marked as different, then the one or two billion yellow minifigures in circulation were not just bodies, but also clearly raced as *normative, unmarked, white bodies*" (2014, p. 317).

be mobilized in play, creating a dynamic bridge between the canonicity of iconic fictional characters and the possibility of character toys to generate an infinite multiplicity of possible playful narratives. At the same time, the fixity of such significations constitutes an implicitly normative playscript designed to align play with the underlying logics of the source narratives, potentially infusing an element of dis-play into play.

LEGO character toys, that is, not only invite play but more precisely invite *transmedia* play as their very material presence signifies archetypal identities designed to be mobilized within specific roles derived from the narrative logic of the source texts. Thus, while these character toys are not filmic in nature, their embodied characterization concretizes, crystallizes, and canonizes archetypes that are inextricably connected to their filmic origins.

Vehicles

As mobile, playable toys constructed in the same manner as other LEGO structures, LEGO *Star Wars* vehicles are essentially hybrids of character toys and playsets. Yet, they also exhibit a distinctive kind of materially-oriented world-building typical of toy versions of technologically-rich military epics. The action-centric transmedia toys of the 1980s often offered mobile narratives of warfare predicated less on contested territories and more on repeated skirmishes between two perpetually opposed sides identified by their distinctive war matériel (and the collectible toy versions thereof). This promoted a knowledge economy in which fans gained cultural capital by mastering information about the distinctive capacities of the vehicles and weapons in the story (hence the importance of compendiums and encyclopedias in offering paratextual accounts of the material culture of the storyworld). Collectible war machines, which make up a substantial percentage of LEGO *Star Wars* sets, thereby construct an ideology of material toys as war matériel (munitions) in which the moral dualism that permeates *Star Wars* storytelling is manifest in differential relationships to military technology.

Star Wars already has a distinctively technological aesthetic featuring recognizable, collectible spaceships that clearly distinguish the opposing factions. As reported by Ryan Lambie (2018), *Star Wars* model designers pioneered a form of model building designed to create the feeling of a “used future,” a science-fictional world in which technology shows age and wear. This aesthetic was produced through literal bricolage—a process of *kitbashing* in which parts from other models were scrounged and repurposed as seemingly purposeful bits of external detailing known as *greebles*, which are particularly important to the AFOL community. As LEGO is a medium that offers kits specially designed to promote kitbashing, greebling can become highly valued as a sign of creative complexity.

Similarly, *Star Wars* operates according to a clear technological ethos that positions its aesthetic of used futures within a broader ideology that values bricolage over industrial production. The transmedia world of *Star Wars* is a world with an implicit moral order that governs the kind of sagas that unfold. In this universe, mass production is linked with oppressive regimes like the Empire and its ilk (including the more corporate Trade Federation), whereas bricolage is the tool of scrappy, creative survivors trying to eke out an existence on the margins, including marginalized bricoleur races like the Jawas. In a telling pattern, the unlikely heroes of all three *Star Wars* trilogies are bricoleurs-turned-warriors whose epic adventures are precipitated by serendipitous encounters brought about by *scavenging* (Luke and Rey both rescue droids with connections to the Rebellion/Resistance; Anakin's hodgepodge podracer becomes the only way for two stranded Jedi to acquire a crucial spaceship part).

And not only does literal bricolage situate these heroes within the material culture of the rebellion, but it also serves as a material metaphor for the values of democracy and diversity[12] that are the implied goals of such rebellions. Whereas mass production threatens to standardize the universe under a single totalitarian regime, *Star Wars* portrays bricolage as freeing the margins of society to creatively reassemble the technological detritus of that regime's violence into the material of active resistance (a similar narrative of resistance operates in *The LEGO Movie*; see Chapter 6). When, for example, Han lovingly maintains his greeble-laden but fast *Millennium Falcon* by making "a lot of special modifications," his technological/affective labor imbues a sense of hard-won belonging into his itinerant, fringe existence. The creative freedom of bricolage becomes both metaphor and means for achieving personal or political freedom.

[12] Although the films do not break from the homogenous casting of their eras until the third trilogy, the original film represents diversity through characters marginalized in other ways—a farmboy (Luke), smuggler (Han), rebel princess (Leia), alien (Chewbacca), and two droids (R2-D2 and C-3PO).

Like the material culture of *Star Wars*, LEGO has become a transmedia world whose very material substructure implies both the construction of war matériel and the deployment of such material within scripted dramas of technologized warfare. Going well beyond the mere fact of LEGO having gun elements, the prevalence of LEGO war machines emphasize the potential of the entire medium to be at any moment militarized, a narrative that achieves its fullest expression when the citizens in *The LEGO Movie* militarize the signifiers of their capitalist oppression in an act of defiant bricolage (even transforming an ice cream cone into a cannon; see Lee 2019).

These themes all meet in LEGO *Star Wars*: *The Freemaker Adventures* (2016–2017), an animated television series about three siblings who run a small scavenging and spaceship repair business becoming embroiled in the galactic conflict. These original minifigure characters take the themes of scavenging and vehicle repair that represent creative freedom in the *Star Wars* universe and place them front and center, drawing out construction as a central ideology in the *Star Wars*. At the same time, this series layers this ideology onto its representation of LEGO construction, drawing a clear linkage between the ethos of creative freedom in *Star Wars* and material play with LEGO *Star Wars* toys.

Thus, when players perform the material bricolage of constructing LEGO *Star Wars* vehicles, their performance situates them within the technologized ethos of the *Star Wars* universe. Consequently, the performance of construction play becomes much more than a material practice, functioning at once as dramatic and transmedia play. Construction play can be dramatized in this way because the scripting of LEGO already emphasizes and ethicizes the process of bricolage. While in one sense LEGO instructions invite players to become laborers merely assembling already-designed sets as if they were part of an apparatus of mass production, the LEGO ethos resists this overly mechanical narrative.

While the building instructions function like blueprints, specifying exactly what part goes where, they also procedurally teach a mechanical skill akin to the tinkering that *Star Wars* positions as a form of creative freedom that many of its paratexts, such as the *LEGO Star Wars: Build Your Own Adventure* guide (Fig. 5.2), explicitly encourage. Implicit in the fixed procedure of constructing according to the instructions, therefore, is a ***procedural rhetoric*** (Bogost 2007) that encourages even the most mechanical builders to build fluency in the underlying principles of LEGO design, including both the grammar of LEGO connection and

Fig. 5.2 A sample page from *LEGO Star Wars: Build Your Own Adventure* (2016) by Daniel Lipkowitz, depicting strategies for building elements of a Jawa Junkyard from found elements. This emphasis on bricolage is also highlighted in the thematic content, which emphasizes the 'recycling' of elements. The pedagogical function of this guide is highlighted in the blueprint-like captions and exploded diagrams that explain the logic behind these designs. For instance, the center panel on "Trash Chompers" exposes a cleverly hidden set of building techniques inside the Recycler through an exploded diagram showing how to use Technic elements to invert the directionality of the LEGO studs

the symbolic vocabulary of LEGO elements. And, as the grammar and vocabulary of the LEGO medium has become thoroughly entangled with the symbolic material culture of *Star Wars*, this is as much an ideological engagement as a technical one.

LEGO story toys do not so much describe this technological ethos as playfully enact it. Even when players replicate the exact models displayed in the building instructions, they do so by piecing together symbolic elements that reconstruct *Star Wars* in toy form, thoroughly implicating players in its narrative logic, an oppositional, dualistic, and technologized ideological construction. And although following the building instructions certainly implicates players in this ideology, deviating from the instructions is not necessarily freedom. As both an explicit and implicit playscript, the LEGO building instructions condition construction play whenever it follows the general design principles these instructions teach.

As militarized LEGO vehicles become vehicles for militarizing LEGO, they infuse LEGO with an implicit bent toward technologized and militarized play. This thereby infuses the medium with a cinematic bent that situates transmedia play within familiar filmic adventure genres. Militarized LEGO vehicles are not only mobile agents for dramatic play—instead, they also serve as *stages* that script narrative possibilities in a similar manner to *playsets*.

Playsets

For most story toys, playsets are relatively static backgrounds or stages for character- or vehicle-centric narrative play. For LEGO, however, constructible playsets take a starring role; as Wolf boldly argues, "Combining the best aspects of building sets and playsets, LEGO has become the most versatile worldbuilding toy available" (2012, pp. 138–139). Rather than seek to immerse players in a fictional world, LEGO story toys invite players to build real material worlds that, like film sets, stage the fictional worlds they represent. Intertwining construction play and dramatic play, LEGO playsets weave the narrative logic of transmedia stories into the signifying structure of its miniature toy stages. Similar to how toy theaters (see Chapter 3) often explicitly adapted famous stage plays, these playsets construct a theatrical world for reproducing familiar media narratives. Yet, more than mere backdrops for character-driven narrative play, these constructed playsets creatively reinterpret fictional worlds as spaces of scripted int(e)ractability, evocative spaces that derive vast storytelling potential from the immobile signifiers they place on dis-play.

LEGO story toys build material worlds that remediate fictional storyworlds less as static environments to be immersed in and more as dynamic environments to be perpetually reconstructed. Whereas most action figures appear ready to burst into action straight from their clear packaging, LEGO sells storytelling with some assembly required. And whereas many storytelling media aim to establish immersion as quickly as possible, LEGO sells a slow build. In other words, one cannot simply be thrown[13] into LEGO worlds, but must painstakingly construct them.

Unlike many character-centric media, the slow build invites players to dwell deeply in the design philosophy of the story toy. This is also a kind of immersion, but rather than immersing an audience in a world that frames the introduction of subsequent details, it immerses players in the building of a world from individual details. Transmedia play in LEGO is therefore predicated on materially reconstructing the toy world according to a designed synthesis between the LEGO medium and the prevailing logic of a transmedia world. This alignment of the toy world with the material culture of the storyworld leverages construction play to create settings whose symbolic order structures dramatic play.

Yet, while world-building potential is what most distinguishes LEGO story toys from other licensed toys, LEGO playsets exhibit a fairly conservative approach to adaptation, often straightforwardly reproducing canonical settings and scenes. Like the theatrically-oriented stages discussed in Chapter 3, these structures architecturally script dramatic play to reinforce the canonical play displayed on the product packaging. Take, for example, the final instruction step from the Hoth Wampa Cave instruction booklet (Fig. 5.3), which includes two non-LEGO graphics—a child's hand and motion arrow—that transform a building instruction into a playing instruction. This step shows how to operate a mechanism for launching a lightsaber to recreate the film scene where Luke uses the Force to retrieve his weapon from the snowbank.

What is notable here is not that the instructions ask players to recreate this scene, but that LEGO includes a mechanical device for mediating this motion. Even assuming a consistent launch, this device is vastly inferior to even a child's hands in terms of simulating the movement of the film, not least because minifigure Luke has no way of catching the lightsaber. Thus, although this mechanism actually makes it more difficult to recreate the scene as the intervention of a non-mechanical Force, it does transform an otherwise generic snow cave into a distinctly *Star Wars* stage.

[13] Philosopher Martin Heidegger (1927/1962) uses this term to indicate how humans come to consciousness in relation to a world that always already precedes them. While mediated worlds don't already 'exist' in quite the same way, the experience of media texts is one of the audience being slowly introduced to a world and story as if they were similarly 'thrown.'

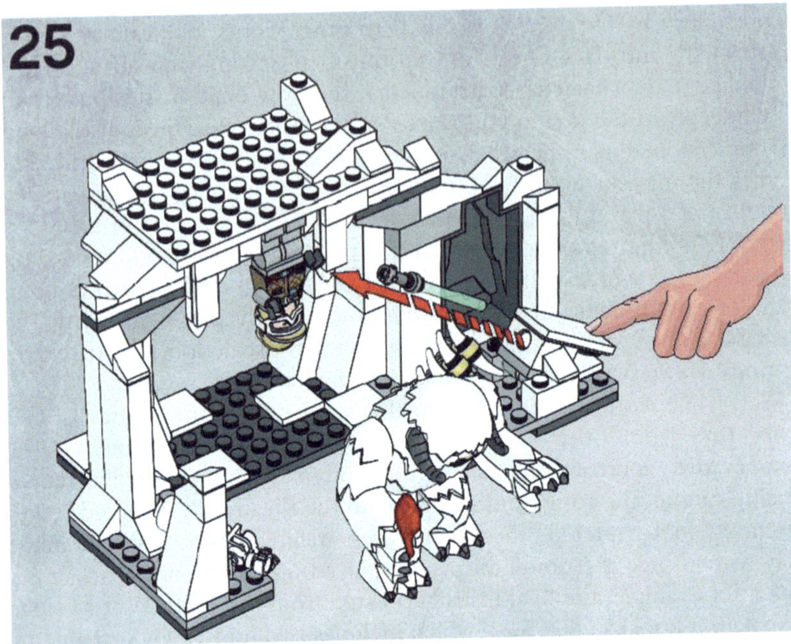

Fig. 5.3 Final step from the LEGO *Star Wars* Hoth Wampa Cave (Set #8089) instructions. The inclusion of the hand and action arrow depicting the lightsaber in mid-launch transition this from a true building instruction to a playing instruction

This world-building operates quite differently from the depictions of the same scene in the film or its novelizations. In such media texts, the narrative of losing and retrieving the lightsaber is a feature of plot rather than of world. The transcendence of worlds beyond narrated events means that audiences are intended to assume that the depicted cave has all the environmental properties typical of any snow cave. Luke demonstrates resourcefulness because he must assess the full complexity of the environment to execute an innovative escape plan. The LEGO cave, in contrast, is one in which the physical environment already predicts the exigencies that drive the plot. In other words, the playset is more akin to the film set than the fictional worlds such sets are used to convey. Such designs draw back the curtain, exposing their inner workings to provide a consistent reminder of the artistry required to cue the imagination of fictional worlds. The signifying strategy of the playset and its instructions thereby aims to simultaneously represent two things—the fictional world as an expansive environment (*Star Wars* as a stage for telling infinite stories) and the constructed world as a mechanism for replicating canonical plotlines (*Star Wars* as a stage fated for telling a predetermined story).

In this way, all three aspects of LEGO story toys—character toys, vehicles, and playsets—script play as a constitutive tension of putting elements of dis-play into play. As these elements are derived from canonical transmedia narratives, the world-building potential of the story toy follows a decidedly filmic logic. Taken together, the hundreds of products that make up this licensed line promote the perpetual extension of the universe while also reinforcing simplistic narrative logics, most notably how the moral dualism of the *Star Wars* plays out in the perpetual opposition of clearly marked factions. Like the ever-expanding universe of *Star Wars* stories, this LEGO product line encourages players to endlessly construct new worlds and tell new stories—while also providing implicit playscripts that encourage such stories to reproduce the familiar narrative tropes that define the *Star Wars* franchise.

Play on Dis-Play: Transmedia LEGO Paratexts

As we have seen, the underlying irony of LEGO story toys is that toys that are arguably most open to reconstructing worlds of any transmedia form are simultaneously wedded to playscripts that encourage reenacting canonical plotlines. This materially interact-able medium of play, that is, relies heavily on the intractable dis-play of fixed signifiers that resist narrative improvisation. Conversely, it turns out that LEGO paratexts in more intractable media often offer more creative, playful, and divergent scripts.

This creates a strange reversal in which non-interactive media follow fixed plots that nonetheless cue transmedia worlds that imply infinite alternative possibilities (the optative mode) whereas the LEGO story toy allows for playing out infinite alternative plots while implying only a single canonical script. Most importantly, transmedia LEGO paratexts aim to draw out the potential for creative world-building with collectible story toys that the toys themselves often fail to embrace. This section explores how two influential paratexts—the LEGO *Star Wars* videogames and television serials—leverage their media-specificity to construct a particularly playful ideology of toy play, speaking to the playful spirit of the material toy to directly counterbalance its more rigid scripting.

This paradox is most clearly expressed in the LEGO *Star Wars* videogame franchise, an odd amalgam of relatively linear, scripted gameplay and whimsical, parodic cutscenes that reenact *Star Wars* scenes in comic pantomime.[14] Like *LEGO Dimensions* (see Chapter 4), these videogames are puzzle-platformers with linear level progression that force players to closely mimic the canonical film plots. Rather than playing out alternative plotlines, these videogames leverage their interactive medium to add contingency to the plot—players can advance the canonical narrative or not advance at all. At the same time, the instantaneous respawning of 'killed' characters transforms this videogame logic into a toy logic.

Typically, the threat of failure in videogame logic produces tension analogous to the threat of failure in *optative* (Miller 2008) narratives—successes feel rewarding because of their contingency. Consequently, most videogame design aims to produce a state of *flow*[15] that maintains an energizing tension by balancing gameplay difficulty against player skill. Contradicting this principle, however, in the toy-centric LEGO *Star Wars* games, death is *never* final—even without restoring from a save, characters immediately respawn with the only consequence being a deduction in the total number of studs accumulated (studs also respawn, so there is no permanent loss here either). As one of the game designers explains, this virtual immortality was an intentional design decision inspired by "learning that children play best in the real world when they feel safe" (qtd. in Newman and Simons 2011, p. 245).[16] In this design philosophy, obstacles are used to enhance a sense of narrative exploration without overtaxing the player's ability to overcome them.

This dynamic replaces the contingency of success/failure characteristic of videogames with the contingency of endlessly repeatable play characteristic of story toys. At the same time, it also replaces the contingency characteristic of linear narratives (the optative threat of narrative failure,

[14] This sharply contrasts with the *Lord of the Rings* game (Fig. 5.1), which layers similar visual humor atop canonical narrative from the films.

[15] Videogame studies, such as Jesper Juul's (2005) *Half-Real*, often cite the concept of flow, developed by psychologist Mihaly Csikszentmihalyi to describe what makes certain activities intrinsically motivating.

[16] Although I do not know specifically where the designers encountered this insight, it is certainly reminiscent of the *attachment theory* discussed in Chapter 6.

including character death) with the contingency characteristic of transmedia worlds (the immortality granted to ostensibly mortal characters who have infinite potential to be redeployed in further stories). Repetition, after all, is a central feature of toy-based storytelling. Thus, Stephen Kline writes that "The stories most children tell in their character-toy play are highly episodic" (1993, p. 339). Embracing contingency without finality, these toy-centric videogames offer worlds of exploratory play within the confines of the canonical plot.

This refusal of finality aims to speak to fans for whom *Star Wars* media represented an endlessly expansive world that transcended its canonical plot. As Robert Buerkle notes, "*Star Wars* was not just immersion in a trilogy of films, but rather a meta-experience: the larger experience of how kids revisited, expanded on, and performed the trilogy. It was an experience of youthful fandom. But more than anything else, *Star Wars* was the *toys*" (2014, p. 132), which facilitated limitless play in a media culture where feature films were not available on demand. Thus, while these games rather straightforwardly adapt the canonical plot, their deeper significance is to evoke what Buerkle calls *playset nostalgia*, a nostalgic reimagining of the idealized toy play of childhood. To accomplish this, "The game doesn't position the player as Luke Skywalker; it positions the player as a child with a Luke Skywalker action figure. It recreates the playset experience" (p. 138). In other words, playset nostalgia is a doubled mediation in which the game allows players to reenact using toys to reenact the story—or, more pointedly, to recall the affective intensity of such transmedia play.

And even though LEGO and *Star Wars* did not collaborate until well after the era these games nostalgically reimagine, the media-specificity of LEGO contributes to the intensity of this nostalgia, which depends on the ideological construction of child's play being somehow more pure, innocent, or affective than cynical adult play. After all, "as a signifier for childhood and toy play" (p. 148), LEGO already has a nostalgic aura. Furthermore, LEGO offers a cartoonish abstraction that reclaims the "eyes of a child" (p. 121) necessary for a nostalgic immersion that can rediscover the playful obsessiveness attributed to the child fan. This works because nostalgia is a feeling constructed in the present—while it idealizes the past, nostalgia is often more powerful when it reconstructs fragments of emotional memory into a creative bricolage than when it accurately remembers historical details.

Engaging this affective bricolage, LEGO enhances nostalgia by infusing canonical *Star Wars* plots with a playful spirit of *parody*, which in the postmodern era, Buerkle argues, "has been repurposed as a rejuvenating process, a way of acknowledging a formula's mythic nature while embracing it as a valid and pleasurable myth" (p. 145). Beneath its ironic detachment, in other words, parody facilitates a kind of nostalgic sincerity, a becoming-childlike that thoroughly reimagines narrative immersion according to an ethos of pleasure rather than fidelity. The videogames reflect this parodic impulse with cut-scenes that replace familiar dialogue with silly noises and mix in a variety of visual gags. Consequently, without representing any actual child's play, these videogames nostalgically evoke the affective intensity and idealized playfulness of child's play.

The parodic character of LEGO *Star Wars* is even more clearly expressed in a series of television specials that aired on the Cartoon Network beginning in 2005, which inaugurated an ongoing string of animated series featuring both adapted and original characters. Deliberately rejecting the seriousness of adult culture in favor of childlike play, these television parodies gleefully subvert the strictures of fidelity that typically constrain *Star Wars* adaptations. The LEGO *Star Wars* series are notable outliers as authorized narratives that blatantly disrupt the continuity of the Lucasfilm-era source narratives.[17] This includes inconsistent characterization (Emperor Palpatine as silly rather than sinister, and bipolar rather than deceptive in his dual roles as Sith Lord and Senator), anachronisms (Darth Vader and Darth Maul competing for the Emperor's affections, a young Han Solo meeting Clone Wars Yoda, and even an entire 2018 series in which a battle droid wanders across the *Star Wars* timeline), and breaking the fourth wall (Vader arguing with George Lucas [Fig. 5.4], Luke Skywalker being mobbed by fangirls).

Despite all this, these series bear the official *Star Wars* logo as fully authorized licensed texts, suggesting that story toys may canonically transcend canonicity. I believe such a ludicrous level of narrative inconsistency is tolerated and even welcomed because, unlike other transmedia texts, the toy-centric mediation of the LEGO universe values playfulness over

[17] Although there were many contradictions within the Lucasfilm-era *Star Wars* canon, which operated under a system of differential canonicity in which the films reigned supreme and other media had more leeway for contradiction, there was also a prevailing imperative to try to tell stories that generally fit the canon.

Fig. 5.4 Screenshot from *LEGO Star Wars: The Padawan Menace* (2011), depicting George Lucas directing an overeager Darth Vader to follow his cue properly. Interestingly, although the only distinctly non-LEGO element in this scene is the folded script, the most meaningful scripting comes from the significations embedded in the minifigure character designs

fidelity. Here, the medium truly is the message as the mediating addition of "LEGO" allows LEGO *Star Wars* to transcend typical *Star Wars* canonicity.

And once again this spirit of playfulness orients these serials around a toy logic that embraces a comic immortality drawn from the durable yet mobile nature of the character toy.[18] Like the respawning videogame characters, these serials involve a form of cartoon violence that reflects

[18] Film theorist Alan Cholodenko (2014) argues that animation is characterized by a play of *lifedeath*, an evocative yet deconstructive collision between the inseparable categories of life and death, animation and inanimation, mobility and immobility. That the medium of animation is used to animate inanimate toys whose material design also implies playful animation seems like a pertinent application of this theory.

a world of temporary pain without the permanent loss of death—any character can be reassembled if 'killed' and every ship can be rebuilt if 'destroyed.' This immortality represents the transition from epic to cartoon, from seriousness to parody. That is, the serials perform transmedia play as storytelling that incorporates its retelling into itself, a logic of the 'do-over' that derives from toy and game play rather than narrative (or, if it derives from narrative, from the re-take of film or the repeat performance of the theater).

Transmedia play in LEGO *Star Wars*, therefore, undermines the threat of death that is, unsurprisingly, a central theme of a film series about a galactic war. At the same time, these transmedia paratexts also clearly link the immortality that softens the dramatic tension to the practice of bricolage. This transforms and extends the meaningfulness of bricolage already within the *Star Wars* universe. In the material culture of *Star Wars*, as embodied in the vehicular story toys, bricolage breathes life into a technological world threatened by violence, attrition, and obsolescence. Significant plotlines include the repair of the starships that cybernetically extend human agency and the reassembly of anthropomorphic droids (most notably C-3P0). In *Star Wars*, however, technology is also what most threatens organic life, from its obsessively destructive war machines to the cybernetics that symbolize Vader's lost humanity and Luke's lost innocence. By contrast, LEGO *Star Wars* toys technologize the organic by rendering all minifigure bodies as droid-like assemblages of interchangeable parts. And its media paratexts reinforce the life-giving connotation of bricolage, going beyond the valuation of bricolage as a means of personal and political resistance already evident in *Star Wars* to embrace a toy-centric narrative logic in which the liveliness of transmedia play depends on material toys functioning as possibility spaces for endlessly replayable narratives.

Whereas the design of the story toys emphasizes the fixity of its signifiers to strengthen their connection to canon, these paratexts remind LEGO players that this fixity is perfectly compatible with the mobility of the story toy. Using more linear media to narrate non-linear, parodic play, these series put play on display in an attempt to reintroduce playfulness into the dis-played signifiers of story toys. Thus, these paratextual playscripts recontextualize the embedded playscripts of story toys to restore transmedia play to the playful mobility that its story toys implicitly offer but often struggle to explicitly facilitate.

Constructing Transmedia Worlds

Just as play and dis-play reinforce each other, LEGO *Star Wars* takes shape in the co-constitutive relationship between story toys and media paratexts. The mutual, co-constitutive intersection of play and dis-play collapses the irony of playful-yet-scripted story toys straightforwardly retelling the canonical story while scripted-yet-playful transmedia paratexts playfully reimagine the canonical storyworld.

Dis-play enables story toys to evoke transmedia worlds by fixedly signifying their material culture. Assuming that play is so thoroughly embedded in the toy system that it need not be articulated, LEGO story toys emphasize dis-play because this connection needs to be explicitly forged. At the same time, the playful paratexts leverage their media-specific modes of display to present toy-centric narratives that authorize players to reimagine the narrative universe in playful or parodic ways. These media paratexts emphasize play over dis-play because their linear textuality dis-plays enough symbolic continuity with the source text that they can freely play with it without severing their connection to the represented universe.

The toys and paratexts emphasize the media-specific characteristics of each other's modality—not to negate their own modality but to forge a transmedia connection. Rather than a single storyworld being incidentally explored in multiple media, the transmedia world of LEGO *Star Wars* only makes sense as a media ecology thoroughly dependent on the playful synthesis of seemingly contradictory media-specificities.

And yet, this synthesis cannot escape the ever-present potential of play and dis-play to negate each other, as this signifying system exerts pressure on transmedia story toys to transform them into licensed collectibles whose significance is entirely in dis-play. Not an either/or proposition, transmedia play in LEGO an exploration of play *and* dis-play or even play *as* dis-play (and vice versa). Consequently, these material story toys and their attendant paratexts uniquely realize the central tension of participatory media culture, the intertwining of consumption and production. To consume LEGO story toys is to produce transmedia worlds and narratives. And to produce new narratives in LEGO is to consume branded products and their embedded playscripts. Unlike most traditional adaptations, the bricolage of LEGO's transmedia play necessarily reweaves already-significant narratological elements into new narratives that are more or less consistent with the implicit directives given by the embedded playscripts. Combining prop, stage, and script, such story toys mediate mimesis as make-believe in a uniquely tangible way.

LEGO story toys materialize otherwise abstract, imaginative, or *virtual* world-building processes. In such cases, it is not that the material replaces the virtual but rather that the material anchors the virtual in a graspable, int(e)ractable form. The story toy becomes a transmedia totem emblematic of the full imaginative potential of the virtual storyworld. LEGO story toys uniquely mediate fictional worlds by creating material playspaces that are fully-functioning, constructible worlds with embedded playscripts that promote restaging canonical narratives. As worlds unto themselves, these toys persistently promise the potential to disrupt the canonicity of these scripted elements, as their elements of dis-play are always available to the playful performance of narrative bricolage. Synthesizing the contradictory impulses of play and dis-play, these toys transform the relationship between story and storyworld by making the world into a space of transmedia play. This synthesis breaks storyworlds free from being implied by linear narratives, elevating them to tangible, accessible sites of playful performance.

In so doing, story toys become more than just another medium added to the ever-expanding web of licensed media products that extend popular franchises. Story toys transform how players relate to stories by constituting perhaps the most genuinely transmedial media in participatory media culture. Whereas, for instance, a *Star Wars* novel might extend the universe of the *Star Wars* films without directly invoking the filmic nature of the original films, LEGO *Star Wars* toys mediate the specificity of other media. As a signifying system, these toys materially mediate canonical film representations as embedded signifiers for narrative bricolage. And as a non-narrative media oriented around narrative playscripts, story toys operate within complex paratextual relationships with other media, simultaneously scripting their transmedia dis-play according to a cinematic logic distilled from the canonical narratives and scripting their transmedia play according to a playful, parodic logic conveyed in other LEGO media.

The transmedia world of LEGO *Star Wars*, that is, deconstructs and reconstructs already-significant elements of other media into resources for transmedia bricolage. As the toy medium embraces its status as a media toy, these scripted yet open-ended story toys not only extend media narratives but also remediate the act of storytelling itself by reconstructing and reanimating storyworlds within the almost magical animating potential of inanimate toys.

Post-script 5—(De)Humanizing Stormtroopers in *Star Wars* Brickfilms

As embodied performance, most transmedia play is too ephemeral and localized to reach into more public media ecologies. Yet, the transformative potential of such play can be amplified when put on dis-play in other media. Consequently, a doubled transmedia play takes place when LEGO transmedia fan productions circulate through unofficial yet public channels such as fan conventions, social media, and photo- or video-sharing platforms. While some of this fan production certainly adopts and extends LEGO's playscripts, fan production can also reimagine stories and storyworlds by transformatively reconfiguring the mobile signifiers of story toys. In the case of LEGO *Star Wars*, a long tradition[19] of fan-produced stop-motion *brickfilms* (also known as *brick flicks*) provide peripheral counter-narratives that explore alternate dimensions of the core transmedia play of both LEGO and *Star Wars*. While much scholarship on fan videos traces the legal and cultural snarls surrounding the appropriation of licensed media by fans and the appropriation of fan production by media companies, this Post-Script considers how several Stormtrooper brickfilms leverage the media-specificity of story toys to re-animate the characterization of Stormtroopers in *Star Wars*.

The defining media-specific feature of brickfilms is their use of genuine stop-motion animation. Unlike the computer-generated animation used in official LEGO *Star Wars* media, stop-motion animation directly draws upon the affordances of the physical medium. As Baichtal and Meno explain, "Brick flicks are a natural extension of LEGO play. All kids—and even some adults—move minifigs around as if they're alive and can talk. These videos simply record that story-telling play" (2011, p. 183). Further accentuating the play of life/animateness and death/inanimateness that Alan Cholodenko (2014) argues defines animation, brickfilms depend on the visible contrast implicit in mobilizing static photographs of plastic toys. This has several media-specific effects that play well into the function of Stormtrooper brickfilms. First, within the narrative world of *Star Wars*, this play of life and death fits well with

[19] *Star Wars* brickfilms are almost as old as both *Star Wars* (1977) and brickfilms (1973), with Freddy Escovar's 1980 *Lego Wars* telling a short vaguely *Star Wars* story with a mashup of existing play themes such as Space and Castle.

the object-like life of the dehumanized, mechanized, and interchangeable Stormtrooper characters. Second, within the broader media ecology of *Star Wars* fandom, stop-motion emphasizes a fannish affective investment by visually emphasizing the material presence of the fan collection and the material labor of the brickfilmer.[20]

Interestingly, both forms of fan investment are often encouraged by the dual controlling franchises of LEGO and *Star Wars*. While the turbulent relationship between amateur creators and Lucasfilm is beyond the scope of this Post-Script, LEGO has often encouraged brickfilmers by periodically releasing "movie maker" products, including the Steven Spielberg Moviemaker set (Set #1349), which included a constructible film set as well as a camera, software, and an extensive guide for creating stop-motion videos. In an introductory note to this guide, Spielberg himself links toy play and public-facing fan production, saying "You may not even realize that when playing with your own toys, you were already telling stories. Now you have a chance to share with your friends the stories you have been telling all these years." Spielberg's comments ostensibly advertise this set as mediating a transition from consumer into producer.

Countering this promise of achieving status as a producer, stop-motion is a medium with a clear DIY aesthetic that marks this production as unofficial.[21] And the reliance of stop-motion animation on consumable LEGO products only reinforces its consumer-oriented production. In other words, LEGO does not so much promote consumers to producers but rather commodifies the kinds of production that emerge with and within LEGO consumption.,

At the same time, the physicality of stop-motion brickfilms means that these fan productions are more directly reflective of the LEGO medium than LEGO's official productions. By contrast, while LEGO's media productions are centered on the physical toy, they mediate the *idea* of the physical toys more than their reality. Outside of its feature films, LEGO's animated minifigures bend and move in more anthropomorphic ways than physical minifigures (see Fig. 5.4), representing the imaginative vision of LEGO players rather than the physical reality of LEGO toys.

[20] Here, I follow the precedent set by Einwächter and Simon in calling these creators brickfilmers instead of fan vidders because "It should also be noted that while brickfilmers are often Lego fans, not all brickfilms are necessarily Lego fan films" (2017, §6.5).

[21] When four brickfilms were officially sanctioned to be included in *The LEGO Movie* (Fig. 1.3), it was because of rather than in spite of this DIY aesthetic.

In contrast, cleverly accentuating the material constraints of the medium—often to comic effect—is a defining feature of the brickfilm. For instance, one visual gag in the *Lego Star Wars Special* by FlapJack Films (2016) exposes this disjunction. In preparation for an imminent battle, an Imperial commander gives the order to "fire the stud-shooter." Here, the film cuts from stop-motion to traditional film as a human finger presses the button on the stud-shooter, launching a stud that bounces harmlessly off the stationary enemy target. As with the lightsaber-launching mechanism discussed in the chapter, the physical reality of this interactive toy fails to evoke any of the dynamism of *Star Wars*, becoming instead a source of humor. This moment demystifies the idealized animateness of LEGO *Star Wars* and instead stresses the inanimateness of material LEGO (even its interactive mechanisms). Similarly, one could argue that the medium of stop-motion animation itself visually emphasizes inanimateness by allowing viewers to perceive the gaps between static frames even while piecing such frames into moving images (similar to how LEGO reveals its constructedness through the visible seams between constructed elements).

Thus, although brickfilms literally use bricolage to construct filmic narratives, they are perfectly suited to deconstructive storytelling because their media-specificity persistently exposes its constructedness. This allows brickfilms to not only reframe plotlines but also to play with the underlying narrative logic of the franchises themselves.

One common thread in brickfilms is the persistent re-imagining of Stormtroopers in ways that play with generic identities and comic violence of a markedly different tenor than that of official LEGO *Star Wars* media. Whereas LEGO cultivates a logic of toy-based immortality designed to insulate the supposedly "innocent" child from the threat of permanent loss, many brickfilms derive a ludic pleasure in gratuitously or comically exaggerating toy-based violence. While perhaps less sinister than some forms of violent toy play,[22] multiple brickfilms emphasize the grisly, tragicomic haplessness of faceless Stormtroopers resigning themselves to battles they know will lead to their deaths or futilely striving to escape

[22] Research on early Black dolls, for instance, shows that their advertised durability played into a cultural expectation that children would enact real physical violence on these represented Black bodies, a clearly and distressingly racialized alternative to the nurturing play more commonly associated with dolls. For a much more nuanced reading of this history, see Robin Bernstein's (2011) *Racial Innocence*.

this fate. Some of these employ gratuitous violence and gallows humor to satirize the disposability of the Stormtroopers. While this might not seem like the most subversive fan art, it does show that fans often take pleasure in subverting the prevailing logic of their beloved storyworlds, even if this entails exaggerating violent overtones to hyperbolically emphasize the logic of disposability present in the original.

More poignantly, many brickfilms seem particularly intrigued by the possibility of authoring quirky subjectivities atop the Stormtrooper mask. This is a sharp contrast to the morally binarized *Star Wars* films, which only humanize the Stormtrooper Finn by removing his mask and having him renounce his Stormtrooper heritage. Without ever removing the mask, the 2013 and 2015 *Storm Trippin* videos by Aaron Legg (Fig. 5.5) blend comedy and pathos by recounting the mishaps of a faceless yet likable Stormtrooper. The first *Storm Trippin* video consists of a series of comic shorts in which the Stormtrooper protagonist clumsily blunders his way through his conscription exams. Here, the hapless Stormtrooper's incompetence is redeemed as his failure to conform to the dehumanizing militarization of the Empire both individuates and humanizes him. In a

Fig. 5.5 An early scene from *Storm Trippin 2: A New Home*, a stop-motion brickfilm by Aaron Legg, showing the courtship of the Stormtrooper and Jedi couple (Image courtesy of the artist)

surprisingly affective scene for a character with no speech or facial expressions, the Stormtrooper glumly tears up his recruitment poster and sits in sullen silence until his reverie is interrupted by a serendipitous encounter with a similarly hapless Jedi candidate. Leaving behind the system that sought to conscript their labor while rejecting their performance, the Stormtrooper and Jedi walk hand in hand into the sunset as the film ends.

The humanizing force of this narrative becomes intermingled with social critique as the second *Storm Trippin* film picks up this journey as the two young lovers (Fig. 5.5), still clad in the garb of their former affiliations, adventure across the *Star Wars* universe in search of a home. After the first two minutes depicting their courtship, most of the film consists of a series of tragicomic sketches in which circumstances—often tied to the perpetual war of the canonical *Star Wars* narratives—ruin all their humble attempts at homemaking, eventually leaving them homeless beggars. As social casualties of a war that refused their services, the lovers eventually humanize themselves by deciding to find home in each other and, eventually, in establishing a collective for the homeless underclass of the Empire. Beyond the pointed social critique, however, lies a deeper narrative critique. Humanizing the faceless Stormtrooper without removing his mask reveals how the *Star Wars* narratives rely on a dehumanizing symbolic logic—it is the filmic logic as much as the Empire that uses the mask to elide any human identity.[23] This transformative work, in other words, draws out the emotive potential of the character toy to transcend and even transgress its canonical narrative function.

Taking a different approach, the violent and racialized 2010 *BlackStormTrooper* brickfilm series, created by and starring actor Donald Faison, offers a bitter satire of a black-armored Stormtrooper fighting back against his criminalization by the very Empire to which he is conscripted. Before *The Force Awakens* showed audiences John Boyega's young Black face behind the mask, this series drew on the dehumanized figure of the Stormtrooper to reflect on racialization in society. Here, the faceless character toy defamiliarizes race by comically exposing the arbitrariness of a character Stormtrooper overacting violent Black stereotypes simply because his helmet is black. While this is less a commentary on *Star Wars* than on race, this brickfilm shows how creative bricolage

[23] For an insightful discussion of the inconsistent humanization of Stormtroopers in the films, see YouTube commentary "The Stormtrooper Paradox" by the Pop Culture Detective.

can appropriate and transform licensed signifiers to tell new and quite different stories. Precisely by relying on the iconic nature of these signifiers, these two brickfilm series are arguably able to provide more nuanced portrayals of the structural violence surrounding Stormtroopers in their short runtimes than does the entire *Star Wars* film franchise.

Any fan production unfolds amidst a constitutive tension between reinforcing and subverting canonical stories and storyworlds. Yet brickfilms add another layer of significance by also playing out the tension between reinforcing and subverting the embedded playscripts contained in the story toys. Brickfilms always make visible the two layers of mediation—story toys and stop-motion—that distance them from the source text. Moreover, the visible interplay between these layers contributes much of a brickfilm's significance. Both the dehumanizing and humanizing Stormtrooper brickfilms, that is, rely on toy-based storytelling. Thus, the loose narratives of gratuitous violence that dehumanize Stormtroopers depend on the heightened interchangeability of inanimate toy figures. Conversely, the humanizing narratives depend upon the affective intensity of the iconic signifiers of the character toy.

Consequently, whether such productions reinforce or subvert LEGO's embedded playscripts, their transformative work is aimed less at reorienting the original canon and more at establishing a public fan response that plays with the canon. Furthermore, as these responses necessarily draw upon the visible presence of the LEGO toy, they respond as much to the transmedia world of LEGO as the transmedia world of *Star Wars*. In so doing, these brickfilms produce narratives of dis-play that speak to the transformative potential of transmedia play to creatively reassemble the already-significant elements that shape stories and storyworlds.

Works Cited

Aldred, Jessica. 2014. (Un)blocking the transmedial character: Digital abstraction as franchise strategy in Traveller's Tales' LEGO games. In *LEGO studies*, ed. Mark J.P. Wolf, 105–117. New York: Routledge.

Baichtal, John, and Joe Meno. 2011. *The cult of LEGO*. China: No Starch Press.

Baker, Neal. 2014. Middle-earth and LEGO (re)creation. In *LEGO studies*, ed. Mark J.P. Wolf, 40–54. New York: Routledge.

Bogost, Ian. 2007. *Persuasive games: The expressive power of videogames*. Cambridge: The MIT Press.

Buerkle, Robert. 2014. Playset nostalgia: *LEGO Star Wars: The Video Game* and the transgenerational appeal of the LEGO video game franchise. In *LEGO studies*, ed. Mark J.P. Wolf, 118–152. New York: Routledge.

Cholodenko, Alan. 2014. "First principles" of animation. In *Animating film theory*, ed. Karen Beckman. Durham: Duke University Press.

Cross, Gary. 1997. *Kids' stuff: Toys and the changing world of American childhood*. Cambridge: Harvard University Press.

duCille, Ann. 1996. *Skin trade*. Cambridge: Harvard University Press.

Einwächter, Sophie Gwendolyn, and Felix M. Simon. 2017. How digital remix and fan culture helped the Lego comeback. *Transformative Works and Cultures* 25: n.p.

Faison, Donald. 2010. BlackStormTrooper. YouTube. https://www.youtube.com/watch?v=YePE2tloDd4. Accessed 24 April 2020.

FlapJack Films. 2016. Lego Star Wars Special. YouTube. https://www.youtube.com/watch?v=csVqeJodPSk. Accessed 24 April 2020.

Fleming, Dan. 1996. *Powerplay: Toys as popular culture*. Manchester: Manchester University Press.

Geraghty, Lincoln. 2014. *Cult collectors: Nostalgia, fandom, and collecting popular culture*. London: Routledge.

Gray, Jonathan. 2010. *Show sold separately: Promos, spoilers, and other media paratexts*. New York: New York University Press.

Heidegger, Martin. 1962. *Being and time*, trans. John Macquarrie and Edward Robinson. New York: Harper Perennial.

Huizinga, Johan. 1950. *Homo ludens: A study in the play element in culture*. Boston: Beacon Press.

Hunter, Madeleine. 2018. Bric[k]olage: Adaptation as play in *The Lego Movie* (2014). *Adaptation* 11 (3): 273–281.

Hutcheon, Linda. 2013. *A theory of adaptation*, 2nd ed. New York: Routledge.

Jenkins, Henry. 2006. *Convergence culture: Where old and new media collide*. New York: New York University Press.

Johnson, Derek. 2013. *Media franchising: Creative license and collaboration in the culture industries*. New York: New York University Press.

Johnson, Derek. 2014. Figuring identity: Media licensing and the racialization of LEGO bodies. *International Journal of Cultural Studies* 17 (4): 307–325.

Juul, Jesper. 2005. *Half-real: Video games between real rules and fictional worlds*. Cambridge: The MIT Press.

Kant, Immanuel. 1929. *Kant's inaugural dissertation and early writings on space*, trans. John Handyside. Chicago: The Open Court Publishing Company.

Klastrup, Lisbeth, and Susana Tosca. 2004. *International Conference on Cyberworlds* 2004: 409–416.

Kline, Stephen. 1993. *Out of the garden: Toys, TV, and children's culture in the age of marketing*. London: Verso.

Konzack, Lars. 2014. The cultural history of LEGO. In *LEGO studies*, ed. Mark J.P. Wolf, 1–14. New York: Routledge.

Lambie, Ryan. 2018. Greebles: How tiny details make a huge Star Wars universe. *Den of Geek*. https://www.denofgeek.com/movies/greebles-how-tiny-details-make-a-huge-star-wars-universe/. Accessed 19 April 2020.
Landay, Lori. 2014. Myth blocks: How LEGO transmedia configures and remixes mythic structures in the Ninjago and Chima themes. In *LEGO studies*, ed. Mark J.P. Wolf, 55–80. New York: Routledge.
Lee, Jonathan Rey. 2019. Master building and creative vision in *The LEGO Movie*. In *Cultural studies of LEGO*, ed. Rebecca C. Hains and Sharon R. Mazzarella, 149–173. Cham: Palgrave Macmillan.
Legg, Aaron. 2013. Storm Trippin (original). YouTube. https://www.youtube.com/watch?v=7SpNHMwwID4. Accessed 19 April 2020.
Legg, Aaron. 2015. Storm Trippin 2—A new home. YouTube. https://www.youtube.com/watch?v=fTS0ez47FHE. Accessed 19 April 2020.
The LEGO Group. 2019. The greatest battle built since 1999. *LEGO.com*. https://www.lego.com/en-us/aboutus/news/2019/april/the-greatest-battles-built-since-1999-celebrating-20-years-of-lego-star-wars-fandom/. Accessed 19 April 2020.
LEGO *Star Wars* Hoth Wampa cave [booklet]. 2010. Set 8089.
McCloud, Scott. 1994. *Understanding comics: The invisible art*. New York: HarperPerennial.
Miller, Andrew H. 2008. *The burdens of perfection: On ethics and reading in nineteenth-century British literature*. Ithaca: Cornell University Press.
Newman, James, and Iain Simons. 2011. Using the force: *LEGO Star Wars: The Video Game*, intertextuality, narrative, and play. In *New narratives*, ed. Ruth Page and Bronwen Thomas, 239–253. Lincoln: University of Nebraska Press.
Ryan, Marie-Laure. 2017. The aesthetics of proliferation. In *World building: Transmedia, fans, industries*, ed. Marta Boni, 31–46. Amsterdam: Amsterdam University Press.
Seiter, Ellen. 1993. *Sold separately: Children and parents in consumer culture*. New Brunswick: Rutgers University Press.
Steinberg, Marc. 2012. *Anime's media mix: Franchising toys and characters in Japan*. Minneapolis: University of Minnesota Press.
Stewart, Susan. 1984. *On longing: narratives of the miniature, the gigantic, the souvenir, the collection*. Baltimore: The Johns Hopkins University Press.
Walton, Kendall L. 1990. *Mimesis as make-believe: On the foundations of the representational arts*. Cambridge: Harvard University Press.
Wolf, Mark J.P. 2012. *Building imaginary worlds: The theory and history of subcreation*. New York: Routledge.
Wolf, Mark J.P. 2014. Adapting the Death Star into LEGO: The case of LEGO set #10188. In *LEGO studies*, ed. Mark J.P. Wolf, 15–39. New York: Routledge.

CHAPTER 6

Toy Stories: Attachment Play in *The LEGO Movie* and *The LEGO Movie 2*

When the father in *The LEGO Movie* (2014) declares that "LEGO is not a toy," he is doing much more than attributing the complexity of a medium to a "sophisticated interlocking brick system." Instead, he denies the medium's association with childlike play to justify his exertion of authority and ownership over what is quite clearly a toy. Thus, the father's behavior demonstrates how "adults, rather than finding toys trivial, are involved in a sometimes buried, sometimes obvious, struggle with children to keep control over them" (Kuznets 1994, pp. 10–11). Yet, rather than having the son Finn wrest control of the toys from his father, *The LEGO Movie* chooses to instead frame Finn's disruptive incursions into his father's LEGO world as an attachment plea, an implicit call to restore the harmony of the father/son relationship.

Accordingly, this chapter explores how the two LEGO movies characterize LEGO as a medium of ***attachment play***,[1] as I call toy-mediated imaginative play aimed at communicating emotional needs and strengthening relational bonds. Although one might think that such play is dependent almost entirely on the dispositions and actions of the players, the films suggest that a disposition toward attachment resides in the

[1] In this chapter, I use "attachment play" to name emotionally resonant play practices within everyday contexts rather than clinical settings. This term does not seem to have much traction in psychological discourse. However, play therapy is a common method used for therapeutically strengthening attachment bonds, yielding terms such as "Attachment-Based Play Therapy."

J. R. Lee, *Deconstructing LEGO*,
https://doi.org/10.1007/978-3-030-53665-7_6

toy itself—a radical claim that is a strange mix of theoretical argument, advertising message, and paratextual playscript. Denying the neutrality of the LEGO medium—the ideal of traditional communications media being to transmit the message without adding or subtracting anything[2]—these filmic paratexts construct LEGO as an active mediator disposed to promote an ethos of attachment.

To deconstruct this ethos of attachment, this chapter loosely appropriates several more or less interrelated psychological concepts that relate to three main components of attachment play: attachment needs, toy play, and storytelling. The first set of concepts come from *attachment theory*, a psychological field pioneered by John Bowlby and Mary Ainsworth. The basic tenet of attachment theory is that to ensure good emotional development, children need to develop strong attachments to at least one parent. Strong attachments provide both a *secure base* from which children feel they can safely explore and a *safe haven* to which they can return from challenging explorations to find love and acceptance.[3] Although attachment theory aims at describing emotional development, its core focus on exploration allows reasonable extrapolation to the creative exploration at the heart of LEGO. After all, the films show Finn's emotional and creative development to be thoroughly intertwined. Thus, when his father restricts, excludes, and overlooks Finn's creative production, this feels to Finn like a lack of emotional support, which is likely why there is so much anxiety evident in his allegorical storytelling play.

While the first film narrates attachment between father and son, the second echoes aspects of this dynamic in the relationship between Finn and his sister Bianca. This departs from psychological attachment theory, which is only just beginning to explore the possibilities of sibling attachment.[4] This departure is necessary to construct a media theory of

[2] It is impossible for any actual communications medium to achieve this ideal because even if a delivery technology achieved lossless transmission of sensory information, the media system would still frame the transmission in ways that inflect communication.

[3] Bowlby discusses the *secure base* in a book of the same name, although he attributes the concept to his student Mary Ainsworth in the dedication. Bowlby (1969) uses the term "haven of safety" a few times in *Attachment and Loss, Vol. 1*, but the concept of the *safe haven* is more central to recent extensions of attachment theory.

[4] A quick glance at some of this emerging research seems to indicate that a theory of sibling attachment would not be exactly analogous to parental attachment. In fact, it seems like sibling attachment is of more interest in circumstances where parental attachment is somehow threatened—divorce, foster care, aging, etc.

attachment play that can more generally account for the interpersonal dynamics at play in everyday toy play, which may reflect the patterns of attachment theory even if they do not constitute attachment relationships in the strict psychological sense (psychology seems to use attachment for particularly formative relationships rather than casual ones). As the films argue, toy-mediated storytelling often aims to strengthen many kinds of relational bonds, including the peer and sibling relationships within which much toy play occurs. Thus, I find that extending the language of attachment may help articulate how LEGO ideologically constructs itself as a medium of attachment play.

For instance, key takeaways from attachment theory involve the interplay of communication and miscommunication in relationships. Attachment theory suggests that in practice, many misbehaviors or conflicts arise from unmet attachment needs. Not only that, such behaviors sometimes constitute miscommunicated attachment pleas, as when "acting out" is a disguised call for attention. This kind of miscommunication provides the central narrative tension in both films as the central resistance and invasion narratives respectively disguise Finn's and Bianca's toy-mediated pleas for relational acceptance.

Crucially, this emphasis on (mis)communication raises the question of how material systems mediate attachment needs. Thus, the toys-to-life narrative that anchors both films depicts LEGO not only as a site of attachment play but as a structuring participant that interjects its own reconciliatory force into the relationship dynamics. In the words of the LEGO Foundation's *Defining Systematic Creativity* report, this suggests that "LEGO bricks are a social tool, fostering connection and collaboration" (2009, p. 12).[5] Although the discipline of psychology does not aim at media-specific analysis, psychological theory and practice provide strong foundations for considering how emotional attachment becomes entangled within the material performances of toy play and storytelling.

Play—and, by extension, toys—have an important role in many child development theories and therapeutic practices. While the specifics may

[5] Similarly, Gary Mankellow, a consultant for the LEGO Serious Play initiative (a program that conducts team-building workshops using LEGO bricks to help facilitate communication), explains that "LEGO lets you talk about an issue when you're unsure of where to start. I keep seeing people that don't know how to communicate find a way to tell their story through LEGO" (qtd. in Bender 2010, p. 222).

vary, many therapeutic practices[6] build on an underlying assumption that deep-seated emotions are engaged in—that is, both invested in and accessed in—the mediated practices of play and storytelling. From a more theoretical and historical perspective, Sigmund Freud's psychoanalytic theory draws a similar parallel between toy play, creative writing (storytelling), and daydreaming as forms of fictive self-expression that reveal unconscious wishes. In "Creative Writers and Daydreaming," Freud (1908/1995) argues that toys become meaningful by being imagined within toy stories, including both narrative play (play crossed with daydreaming) and toys-to-life narratives (play crossed with storytelling).

In this vein, toy stories—including both the toys-to-life narratives of the LEGO movies and the imaginative play of children depicted in the LEGO movies—are often understood as profoundly revealing and even transforming the psyche of the playing child. Similar notions are reflected in a Western narrative tradition of *toy stories*, psychologically or emotionally resonant narratives of toy play that Lois Kuznets notes commonly contain motifs of exploring anxiety about becoming a real self, encountering the uncanny and liminal "secrets of the night," grappling with "the temptations and responsibilities of power," and implying "vital possibilities for human creativity while arousing concomitant anxiety about human competition with the divine" (1994, p. 2). Clearly, in both psychological and cultural traditions, toys and toy stories can conjure powerful subterranean emotions.

Inspired by such theories of attachment, play, and storytelling, I propose here a media theory of emotional attachment, exploring how emotional or relational needs play out within media systems. More specifically, I argue that toys mediate toy stories, and toy stories can in turn mediate emotional attachment by helping players confront and/or communicate attachment needs. LEGO offers a distinctive brand of attachment play for several reasons. With its unique blending of construction and dramatic play and its evocative yet somewhat blank minifigure characters, LEGO offers symbolic bricolage that already inclines toward certain kinds of affective portrayal. Furthermore, as a medium already entangled with messages about its own play, LEGO evokes an emotionally

[6] There are therapeutic practices like Attachment-Based Play Therapy and Parent/Child Interaction Therapy (PCIT) that treat play as a site of strengthening attachment bonds. For storytelling, the field of *narrative therapy* uses storytelling to both uncover and rewrite the stories that shape people's lives.

resonant cultural nostalgia that positions LEGO as an evocative object. More than a communicative tool, LEGO brings with it a culture of play whose many resonances guide its attachment potential.

This evocativeness is only intensified by the fact that LEGO is not merely associated with childhood at large, but particularly with *imagination* and *creativity*, deeply generative practices that speak to the profundity of children's self-expression. Consequently, LEGO provides distinctive possibilities for mediated attachment precisely because it differs significantly from traditional communications media designed to technologically extend the reach of otherwise embodied linguistic communication. Instead, LEGO has its unique way of embedding emotional resonances within evocative objects, a form of material expression that is arguably emotionally intensified by LEGO's symbolic yet non-linguistic character.

As emotional attachment is mediated by these evocative objects, the central ethical problematic of LEGO attachment play lies in how LEGO's potential as a medium of attachment rests on players taking up corporately-authored messages embedded in toy commodities. Reinforcing these messages, the filmic toy stories are simultaneously meditations on attachment and manifestos for the attachment potential of a commercial product. Commercialism is unsurprisingly a prevailing ideology of the films and the substance of nearly all extant scholarship on *The LEGO Movie*. To name a few examples, Matthias Zick Varul (2018) analyzes the film as 'a consumer-capitalist myth,' Dalia Grobovaite (2017) as a branded revisioning of the culture industry, and John Daniel Holloway III (2016) as an expression of 'military consumerism.'

Validating such critiques, consumerism and the establishment of a corporate brand identity infuse each of the four layers of meaning explored in this chapter: the diegetic quest narrative, the psychological frame, the filmic animation, and the authorship of the material toy. Working together to produce a coherent toy story, these four levels thoroughly circumscribe attachment play within the distinctive brand of commodity culture that structures the LEGO medium.

The films initially present themselves as anti-consumerist by stigmatizing how their supposed villains abuse the tactics of capitalist and consumerist culture. In the first film, the Master Builders push back against Lord Business's use of consumerism as a tool of social control. In the second, Wyldstyle, resists Queen Wa'Nabi's attempt to sway the Master Builders using pop music and promises of material gain. Rather than undermine the films' advertising messages, however, these anti-consumerist stances may only make them more potent, as comic irony is one way advertisers diffuse resistance to advertising. Characteristic of self-aware postmodern advertising, the film acknowledges its complicity in ways that appeal to postmodern subjects whose consumption is similarly self-aware (Goggin 2018). And, as Ari Mattes argues of LEGO's "post-ironic media," "irony's sheer ubiquity in LEGO media leads to its flattening, negating irony's potential as bearer of critical meaning" (2019, p. 91). This allows *The LEGO Movie* to undo its anti-capitalist messages and instead achieve "what is every advertiser's secret dream: it produced a myth that manages to negotiate real contradictions through a commodity" (Varul 2018, p. 737), such that "the movie, while seeking to critique capitalism and the effects of mass culture, is actually promoting sameness and is skilfully serving as a perfect example of product placement" (Grobovaite 2017, p. 59). Indeed, as capitalism often commodifies potentially anti-capitalist sentiments like creativity and freedom, this paradox only makes the bright, exuberant, consumerist spectacle of the animated toys more appealing.

As toy stories, the films script attachment play by performing a narrative self-branding that defines the potentiality of LEGO for mediating attachment. Whereas psychological theories might consider how children invest emotional significance in objects, this chapter deconstructs how an ideology of attachment becomes invested in this symbolic, consumer medium. It does this by deconstructing how the two LEGO filmic toy stories position LEGO itself in the role of emotional mediator. As a medium of bricolage, the emotive potential of LEGO draws upon a predisposition to attachment already present in its elements and, more importantly, in a connective system already oriented around the literal *attachment* of physical pieces. In this sense, LEGO functions at once as a site of attachment play and a material metaphor for the underlying concept of attachment itself. As this ideologically-laden system comes to stand for the attachment it conveys, LEGO conjures and script the emotions it conveys. Consequently, these toy stories function as paratextual guides for scripting an attachment play in which children are encouraged to weave their own toy stories with and within the LEGO medium.

To do this, the films nest several layers of meaning within each other. The surface quest narrative, the subject of the first section of this chapter, reframes the quest genre around an attachment plot starring cyborg toy bodies that call into question the distinction between self and world. Second, the revelation of a playing child as the implicit author of the toy story contributes a layer of creative daydreaming in which material toys mediate emotional attachment needs. Yet, whereas the films imply that this is the final layer of meaning, they also reveal a self-referential third level in which the filmic medium and corporate authorship of LEGO provide an otherworldly intervention into attachment play, positioning the LEGO ethos as an active player in this domestic drama. Consequently, the psychological potential of the story toy is thoroughly implicated in the circulatory system of branded commodities, suggesting that this narrative of attachment is as much about brand identification as about interpersonal relationships. In other words, the LEGO in the film not only facilitates the mediation of attachment but problematically redirects attachment toward its own emotionally resonant toys.

Attachment Quests

Although the psychological concept of attachment is more personal and interpersonal, these toy stories use epic storytelling to allegorize attachment-like collectivism. Even before the twists in both films reveal the real-world frames from which the animated toy stories become instances of attachment play (as discussed in the following sections), the animated quests prefigure the importance of attachment by adhering to a consistently collectivist ethos. It is not too much to say that both animated toy stories are essentially *attachment quests*, narratives in which the fate of the world depends on reconciling fractured communities. To unravel how these quests situate LEGO as a medium of attachment play, this section explores the central themes of the two plots and how the figures of Emmet and Queen Watevra Wa'Nabi speak to the attachment potential of the medium as respectively a hero and being of connectivity.

The LEGO Movie stars the cheery yet hapless construction worker Emmet (Fig. 6.1), a relentlessly friendly everyman destined to become a hero of connectivity and resolve the film's attachment quest. Emmet is introduced naively conforming to the 'instructions'[7] that organize his dystopic society. When he accidentally stumbles upon the legendary

[7] Citizens in Bricksburg are given small LEGO plate elements printed with pictorial instructions similar to LEGO instructions to instruct them on appropriate behavior. Emmet is introduced starting his day by following the "instructions to fit in, have everyone

Fig. 6.1 Screenshot from *The LEGO Movie* introducing Emmet as a domestic, everyman hero. Drawing out the dystopic subplot of the film, Emmet's exaggeratedly dynamic pose and cheerful demeanor contrast with the blandness and conformity of his home, which is full of generic markers of consumption, such as the poster labeled "Fast Car" on the rightmost wall

Piece of Resistance, Emmet becomes embroiled in a struggle between the underground resistance of Master Builders and the authoritarian Lord Business. Somewhat surprisingly, the film subverts genre expectations[8] by refusing to allow the individualism of the freedom fighters to straightforwardly triumph over authoritarianism.

Instead, building on the socializing tendencies of the Town Plan and the emotional openness of LEGO Friends, Emmet becomes a hero of connectivity. Without winning a single significant battle, Emmet leads the

like you, and always be happy!" Larger paper instructions also appear in a few places throughout the film, including Emmet's worksite.

[8]As film guru and story consultant Robert McKee and Bass El-Wakil (2014) explains, as the adventures play out across "all these worlds that sort of relate to all the sub-genres of action," the film becomes a "satire of something that's been around for quite a while now: the hero's journey, the quest myth." Here, McKee refers to Joseph Campbell's (2008) influential work on myth, *The Hero with a Thousand Faces*, in which he argues that there is a common structural pattern behind hero narratives in different eras and cultures. However, even within Campbell's perspective, modernity—especially its focus on individualism—generally subverts the hero's journey even while it continues to invoke many of its elements. Incidentally, Campbell's theory draws heavily on psychoanalytic theory (especially Freud and Jung) and strongly influenced the development of *Star Wars*, whose relationship to LEGO is discussed in Chapter 5.

resistance by teaching the Master Builders to cooperate, inspiring Wyldstyle, to lead the ordinary citizens to turn their collective creativity toward resistance (see Lee 2019). Further subverting genre expectations that assume militant resistance, the climax of the film is a moment of reconciliation in which the unlikely and atypical hero Emmet invites Lord Business back into the collective. Although Emmet had always been ostracized for so unabashedly vocalizing his attachment needs, Emmet's childlike invitation offers a safe haven to the insecure adult in a surprising inversion of the typical attachment relationship.

The first film ends with a sudden jolt to this newfound harmony: to Finn's deeply ironic horror, his father comments that the natural extension of Emmet's collectivist logic is that Finn's sister Bianca be allowed to join their play. Immediately thereafter, a group of multicolored[9] Duplo characters crash the celebration, announcing in a babyish voice: "we are from the Planet Duplon ... we are here to destroy you." *The LEGO Movie 2* (2019) picks up from this moment to explore the evolving aftermath of Bianca's contested integration into Finn's play.

After a brief encounter in which the invaders consume the LEGO heart Emmet constructs as a peace offering, *The LEGO Movie 2* skips ahead several years to a war-torn, post-apocalyptic future that attests to the ongoing sibling battles of Finn rejecting Bianca's incursions into his playspace. Whereas Emmet and Wyldstyle became champions of collectivism in the first film, in the sequel they fall into a more traditionally conflict-oriented model of resistance. As an abducted Wyldstyle resists Queen Wa'Nabi's attempts to sway the Master Builders, Emmet resolves to toughen himself up to rescue them. Influenced by his adventuring companion Rex Dangervest's toxic masculinity, Emmet turns to a decidedly anti-collectivist act of destruction.

The twist comes when it is revealed that the Queen is genuinely "not evil." This time, saving the world requires that the would-be protagonists recognize that they have been precipitating the violent conflict and instead reciprocate the attachment pleas underlying the purported 'invasion.' In this more inward journey, Wyldstyle renounces her feigned toughness, catalyzing Emmet's restoration to his role as a hero of connectivity. As Rex Dangervest turns out to be an embittered future version of Emmet who seeks destruction as a misguided attempt to compensate for loss,

[9] The multicolor aesthetic that results from privileging form over color is commonly associated with younger builders.

Emmet must confront himself, renounce his toxic masculinity, and restore his collectivist ethos before he can reconcile the fractured world.

The plots of both attachment quests require the heroes to lay down their swords and instead embrace emotional vulnerability to extend (in the first film) and reciprocate (in the second) invitations to attachment play. Underlying this empowering yet somewhat generic message, however, is a deeper exploration of the mediateness of toy-centric attachment play. The reconciliatory work of the films consistently intertwines social connectivity with collective creativity and collaborative construction. In the first film, all the attachment invitations—Emmet reconciling the Master Builders, Wyldstyle reconciling the citizens, Emmet reconciling with Lord Business—are to *build together* (tellingly, this is also the name of a series of LEGO advertisements featuring fathers and sons bonding over LEGO construction). Similarly, the central symbol of peace in the second film—the LEGO heart that Emmet offers the 'invaders' at the beginning of the film—is a symbol of creative construction.

Connectivity animates the LEGO world against threats of permanence (the glue in the first film) and disassembly (Emmet's attack in the second) that destroy social collectivity by destroying material connectivity. If death for a LEGO character entails reversion to an inanimate object (via glue or disassembly), then life must be movement, mobility, animation. That is, the life of the LEGO character is neither internal nor biological but the quality of circulating within the animating performances of toy play and toy stories.

Thus, the films abound with scenes exploring the connectivity characteristic of LEGO's constitutive interplay of life and death, animation and inanimation, organic and inorganic. A particularly telling example from the first film is the scene where Emmet uses his head as an axle. Here, Emmet's first moment of genuine creativity involves not creatively manipulating the LEGO world (as the other Master Builders do) but humbling himself to the status of an inorganic object. Whereas a mortal human would have to sacrifice humanity to act in this way, Emmet shows how a toy body can object-ify itself while retaining the constitutive harmony between LEGO bodies and worlds. As a hero of connectivity, Emmet allows his body to disappear into the LEGO world the same way his rather generic personality disappears into the collectives he strives to reconcile.

Taking this potent play of embodied animation and inanimation still further, the many forms of the aptly named Queen Watevra Wa'Nabi embody a multiplicity of connective possibilities. A constantly shifting mass of bricks, Wa'Nabi is essentially a personification of the LEGO medium itself in all its interconnective potential. Demonstrating this special status, Wa'Nabi features in the highly unusual Queen Watevra's Build Whatever Box (Set #70825), a uniquely character-centric basic bricks set with instructions for building 15 different incarnations of the Queen. Yet despite personifying a medium, Wa'Nabi is consistently unable to communicate her attachment. Language fails her—not because her speech is error-ridden,[10] but because even her most clear and direct speech is interpreted otherwise. Within the interpretive frame of the film, audiences are encouraged to share in Wyldstyle's, (and, by extension, Finn's) distrust of the Queen's literal and figurative shiftiness. Her genuine linguistic errors are made to sound like Freudian slips revealing a dark secret while her genuine protestations sound too straightforward to be trusted.

Transcending language to rely on her own media-specific form fares no better. Immediately after introducing herself to the Master Builders in the form of a horse (Fig. 6.2a), Wa'Nabi explains that "I can change my form to something else if this makes you uncomfortable" while transforming into a tentacled blob (Fig. 6.2b). Yet this attempt to become "more palatable" only reinforces her alienation. And while it is tempting to simply label Wa'Nabi a bad communicator, her unsuccessful attachment pleas demonstrate more than a supposed incompetence. Alternately, her threatening undertones can be interpreted as attempts to join Finn's more stereotypically masculinized, militarized play. In this reading, there is a reason she feels a tentacled blob might be more palatable than the stereotypically feminized horse-and-princess scene in which she is first encountered. Boys, after all, stereotypically use pretend violence as a means of relationship-building. So, when Wa'Nabi resorts to abduction to advance her marriage plot, the deeper message is that Bianca expects more formal invitations to stereotypically feminized social play to be

[10] Matthias Zick Varul argues that the Queen's "immediately recognizable, emphatically Black voice" combine with her "problems in applying correct semantics" in ways that "play to everyday-racist assumptions regarding a presumed Black language deficiency" (2019, p. 133). This is a solid critique of the film's racial politics that is not undermined by the fact that this "language deficiency" simultaneously places greater emphasis on her media-specific embodied communication.

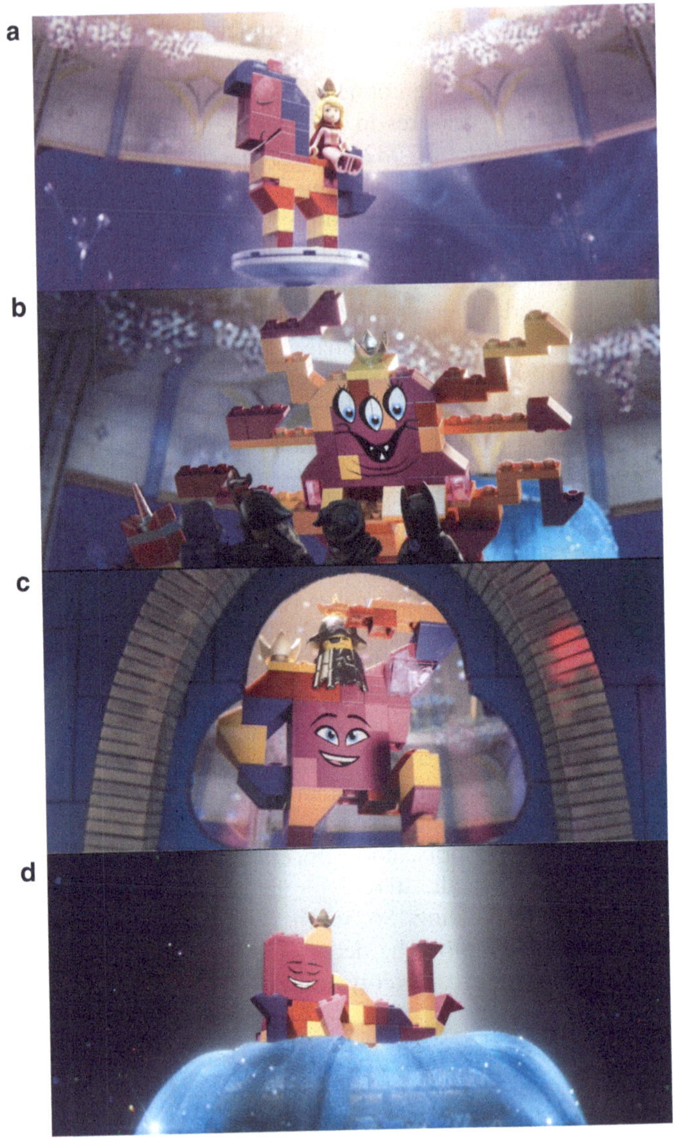

Fig. 6.2 From top to bottom, Queen Watevra Wa'Nabi **a** being introduced as a horse, **b** trying a more "palatable" form, **c** posing as the body to Metalbeard's head, and **d** taking on a rare human form in the finale of "Not Evil"

quickly rebuffed and instead contorts herself to express herself in ways that consistently get misunderstood.

After this inauspicious introduction, Wa'Nabi tries to sway the Master Builders using a "universal language"—musical theater. Singing "I want to shower you with gifts because I'm selfless and sweet," Wa'Nabi woos with audaciously extravagant promises, offering Unikitty infinitely more glitter than she can imagine, Benny "your very own planet," and Batman "half of everything." While such abundance is already persuasive, the deeper promise is of connectivity. Before any material promise is made, Unikitty's eyes light up when Wa'Nabi briefly assumes the form of a cat, and Benny is taken aback by her acknowledging his love of spaceships. Even the recalcitrant Batman is swayed not only by her promise but also by the fact that "this chick gets me." She demonstrates, that is, not only that she has what they materially want, but also that she understands and respects what they genuinely want. Thus, she offers play to Unikitty, construction to Benny, specialness to Batman, and companionship to them all. And these are also implicit promises of the LEGO medium: boundless play, construction, specialness,[11] attachment.

Wa'Nabi not only has and understands what they want, but she also demonstrates that she can *become* what they want. After all, if Emmet is a hero of connectivity, Wa'Nabi is a *being* of connectivity. Thus, while she sings "Not Evil," she performs an entrancing dance that uses her malleability as a personified medium to embody the potentiality she offers. As a being of connectivity, Wa'Nabi's embodied attachment play is most profound in her wooing of the cyborg Metalbeard. Whereas the other characters only see him as stubbornly unhelpful in the first film, she sees what underlies Metalbeard's lament over the loss of his ship, crew, and body—not only fear but also a sense of displaced identity. To restore this, she sings the seemingly suspicious but actually empathic refrain: "a pirate without a ship, that's so cruel—it's like a spider without a web or a queen without a fool." She then promises to restore his lost individual

[11] While this is not one of the core aspects of the prevailing LEGO ideology of *development, imagination, creativity*, it is a central theme of the first film. Emmet initially is thought to be "The Special" when he finds the Piece of Resistance. Yet even after discovering the prophesy to be made up, he embraces his own specialness and uses his ability to see the specialness of others to sway Lord Business to relent in his totalitarian schemes. For LEGO, *development, imagination, creativity* is what make children special, not because they make some children better than others but because they are special characteristics of childhood itself.

and collective identity by offering "a planet that's really a pirate ship and the population: your crew." Even more poignantly, as an embodiment of the LEGO medium, Wa'Nabi performs the intimate, restorative work of physical reattachment by dancing as his absent body, even to the point of fully circumscribing his gaze toward her promised future (Fig. 6.2c). In this moment Wa'Nabi both literally and figuratively connects with Metalbeard, personifying the inherent connective potential of the LEGO medium itself.

True to her name, Watevra Wa'Nabi promises to help her prospective allies become 'whatever you wanna be.' And she does this by personifying a shapeshifting medium, becoming 'whatever you want *her* to be.' While this degree of self-negation is not an ideal foundation for healthy attachment relationships, the films do suggest that some renunciation of individuality is necessary to establish a genuinely collective identity. The minifigure Emmet adapts his personality to reveal the specialness of other minifigure characters and unite them into a social collective. More radically, the brick-built Wa'Nabi shifts her material form to reveal the promise of LEGO construction, personifying how material attachment (material connectivity) promises emotional attachment by being attuned to inner psychic desires (largely for emotional connectivity). Thus, Wa'Nabi not only unites construction and dramatic play in defiance of the gendered dichotomy that typically fragments these forms of play (see Chapter 3)—she also characterizes the constitutive blend of construction and dramatic play as attachment play.

Consequently, the media-specificity of LEGO forges a metaphorical link to attachment that strengthens the already-significant potential of toys to gain emotional resonances in toys-to-life narratives. Any animated character may express attachment in word or deed, but LEGO characters also express attachment in their bodies, which are of their connectable world in ways that are simultaneously earthier and more cybernetic than is typical for humans. To be sure, this material attachment is not how psychologists describe attachment. And yet, in the symbolic world of the toy story, the media-specific promise of LEGO attachment has emotional resonances, promoting the psychologically dubious yet ideologically resonant claim that branded LEGO is a medium ideally suited for attachment play.

Creative Players and Daydreaming

In a bold twist that distinguishes *The LEGO Movie* from many animated children's films, the film's climax takes place in a realistic live-action sequence that reveals that a young boy has been playacting the entire story within the miniature LEGO city built by his father. This frame adds a new layer of significance to the toy-centric attachment quests as the film becomes not only a toy story but also a play story—a meditation on children's emotional investment in narrative play. Drawing upon the psychoanalytic framing of the live-action cellar scene, this section explores the toy-mediated attachment play in the two films: how, in the first, the father discovers Finn's attachment plea in the medium itself and how, in the second, a LEGO heart wordlessly communicates attachment needs. In both cases, the films resist cultural narratives of conflict—the Oedipal contestation of father and son or the gendered contestation of brother and sister—although the latter is more problematic because of how LEGO is complicit in this cultural fragmentation. While these emotional conflicts may run deep, what is most notable about toy-mediated attachment here is how they play out on the plastic surfaces that bear the traces of emotionally-laden play.

Just as the quest narratives subvert genre conventions to promote an ethos of connectivity, this psychological drama rejects the Freudian Oedipal struggle as a foundational desire of child development. Founder of attachment theory John Bowlby explicitly reinterprets the case study Freud used to establish his Oedipal theory to argue that it is "no less plausible" (1973, p. 219) that Little Hans's phobias resulted from fear of attachment loss. Similarly, although the film initially positions Finn's minifigure avatar Emmet against Lord Business's authoritarian regime, the climactic reconciliatory plea shows that Finn refuses to supplant the father.[12]

Furthermore, although Freud suggests that child's play is determined by "a single wish—one that helps in his upbringing—the wish to be big and grown up" (1908/1995, p. 438), Finn instead desires that his father would become more childlike. Here, not only does attachment supersede Oedipal impulses as the film's driving force, the film inverts classical

[12] The mother, whose possession should be at stake in the classic Oedipal triangle, is conspicuously absent in the first film. When she appears in the second film, it is primarily as a disciplinary figure who is not presented in any Oedipal way.

attachment theory by making the child the one to redeem the parent. As McKee notes, "the child recognizes the psychological lacking in the adult character and then relieves the adult character of their immaturity by giving them the positive feeling that they need" (McKee and El-Wakil 2014, n.p.). When Finn *acts out* by transgressing his father's rules, he does so to *act out* a speculative toy-mediated fantasy of more attached play.

Shortly after switching to live-action, the film visually and spatially outlines the stakes of this central attachment relationship by adopting the perspective of the child gazing upward at the backlit figure of his father—known as "The Man Upstairs" in the LEGO world—descending the stairs. Although this low-angle shot lends the father a certain aura of authority, his descent simultaneously brings the father into the cellar, which according to Gaston Bachelard serves as a powerful metaphor of the subconscious (1958/1994, p. 147). This cellar—and the LEGO worlds it contains—is the contested space of the struggle between the ludic imagination of the child and the controlled reality of the father, two paralleled but opposed fantasies. Spatializing the live-action frame in this way suggests that this ostensible move toward a more realistic space is simultaneously a move into a space of deep psychical contestation. The cellar, that is, is both where toy stories find material expression and where material toys obtain psychic significance.

At the same time, relegation to the cellar also threatens toy-mediated attachment pleas that aim to be more than wish-fulfillment fantasies but to genuinely communicate. The father's regulation of the playspace in the first film and the mother's threat to put all the LEGO into storage if the siblings cannot get along in the second film alike threaten to eliminate the potential of the LEGO fantasy to actually engage others.

Whereas a fantasy can take place in isolation, the toy-mediated fantasies of attachment play require circulation to impact relationships. Indeed, in the second film, isolation is what warps the LEGO fantasy. As Finn plays alone in the cellar, stubbornly resisting his sister's attempt to join his play, his isolationism undermines his initial collectivist ethos. As alter egos Emmet and Rex play out Finn's internal conflicts, Rex points out "This isn't even happening. It's all just the expression of the death of imagination in the subconscious of an adolescent." Indeed, Rex's very existence allegorizes the embittering and self-perpetuating paradox of isolation—accidentally lost under the dryer, the pain of watching years of unattainable play transforms the hero of connectivity into a bastion of isolationist toxic masculinity. Thus, both films portray isolation—whether externally or self-imposed—as sites of emotional disconnection that need to be restored to attachment.

Counterintuitively, this restoration to attachment depends on critical moments of silent, isolated reflection. When verbal attachment pleas lead only to conflict—specifically, the verbal sparring of the human characters and the allegorical warfare of the LEGO characters—the films show LEGO itself as successfully communicating the miscommunicated attachment pleas.

In the first film, despite being unable to receive Finn's spoken attachment pleas, Finn's father begins to confront their fractured relationship by *reading* traces of Finn's attachment play in the rebuilt cityscape. In a series of close-ups (Fig. 6.3) that represent the camera's most intimate engagement with both the human face and real-world LEGO, Finn's father absorbs Finn's feelings as his gaze slowly traces over Finn's LEGO creations depicting the climax of the adventure, paused and solidified into plastic bricks. This reading continues after the father recognizes himself in the Lord Business minifigure (Fig. 6.3d), as the camera takes on the father's perspective in a series of tracking shots (not pictured) that show the father looking from Finn's troubled face to the 'Do Not Touch' signs that represent paternal control.

And LEGO not only mediates the father's realization but also mediates their subsequent reconciliation. Tellingly, the father opens a dialogue by invoking the mediateness of the fantasy LEGO world asking, "if the construction guy said something to President Business, what would he say?" At this point, the response almost doesn't matter since Finn's father has already entered into the diegetic world as a full participant in Finn's attachment play.[13] Reminiscent of Marshall McLuhan's (2005) maxim *the medium is the message*, here the mediation is the message—establishing a reciprocal communicative link is both the end and means of this LEGO-created attachment plea.

Building upon the participatory gaze, this crucial scene establishes a significant interplay between words and wordlessness that resonates through both films. Although this scene resolves in parallel dialogue

[13] Similarly, a core tenet of Parent-Child Interaction Therapy is that parents need to enter into and participate in children's play in particular ways to cultivate stronger attachments. Discipline is part of such therapy, but overly strict regulation and overly negative communication like the "Do Not Touch" signs are discouraged.

a
b
c
d

Fig. 6.3 Four screenshots depicting the father 'reading' Finn's attachment play. **a** & **b** Initially drawn in by Finn's skill and creativity as he follows the weaponized pipes to, **c** their epicenter in Lord Business's control room, **d** the father begins to grasp the attachment plea when he recognizes himself as Lord Business. Every one of these shots is carefully framed by the LEGO world, placing the viewer inside the site of Finn's mediated attachment plea and suggesting that LEGO is both what separates and connects father and son

scenes, the initial moment of reconciliation takes place in the wordless moment of the father's asynchronous reading of Finn's attachment play. Even more strikingly, the reconciliation in the sequel when Finn finally responds to Bianca's attachment plea is fully wordless. While Emmet and Wyldstyle, have important realizations due to diegetic conversations, Finn and Bianca communicate their deepest attachments almost exclusively through LEGO, especially the pink-and-purple LEGO heart that is Queen Wa'Nabi's true form.

The sequel opens with Emmet constructing this heart as an offer of peace to the alien Duplo characters, who immediately consume the heart and ask for more, setting in motion the film's central misunderstanding. Finn and his minifigure avatars see this deconstruction as an act of violence whereas Bianca partially evolves her dynamic building style from this encounter (as the heart and its many configurations become her central protagonist). Toward the end of the film, after Finn has violently destroyed Bianca's creations and is putting the LEGO in storage as a punishment, he discovers the fragments of the heart and remembers giving it to a young Bianca saying, "it can be whatever you want it to be" (the origin of Watevra Wa'Nabi's name[14]). As Emmet internally narrates his self-realization, Finn wordlessly enters Bianca's room—where she is sitting in sadness rather than anger—and offers her the reconstructed heart. Once again, the medium becomes the message as the heart functions as a wordless, multilayered material signifier of apology, reminiscence, and attachment promise.

After another interlude in the diegetic LEGO world, the viewers are invited to share the mother's perspective as she observes Finn and Bianca playing together and silently rescinds her previous ultimatum to put the LEGO away for good. Here too, this wordless scene conveys implicit meaning as the harmony she observes is interpersonal attachment mediated by their shared LEGO play.[15] It is unclear whether the mother

[14] Thus, Wa'Nabi is born as the personification of the medium. Finn's statement is about the promise of the medium: "it [LEGO] can be whatever you want it to be." Bianca, however, concretizes this sentiment by misinterpreting—or, more precisely, reinterpreting—this gesture as a christening: "it [this LEGO heart] can be Watevra Wa'Nabi."

[15] Interestingly, their play is largely dramatic (and somewhat transmedial). The brief scene seems to indicate that rather than constructing new structures that merge their building styles, they instead restore their independent construction styles as they were and then subsequently create crossover narratives.

is reevaluating her decision that her bickering children needed to be punished, or reevaluating LEGO as a medium of attachment rather than a medium of discord. After all, her punishment was equally ambiguous—relegating their LEGO collections to the storage bins was as much an indictment of the medium as the players. So, while the scene is intended to convey a parenting decision about sibling relationships, this decision is necessarily intertwined with a judgment about the attachment potential of the toy medium. In materially communicating miscommunicated attachment pleas, LEGO proves its worth as a medium of attachment and thereby plays a role in its own salvation.

This is particularly important because the second film also implies that the medium is to some extent complicit in fracturing the sibling bond. While the medium was at stake in the first film, the schism was largely due to the father misconstruing LEGO as static representation rather than connection, clearly placing him at odds with the LEGO ethos. By contrast, the attachment plea in the sequel is predicated on a gendered division in the toys themselves (see Chapter 3). Finn's LEGO is inherently not as compatible with Bianca's as with his father's, as evidenced by the differential gendering of the different LEGO movie product lines (Fig. 6.4). Although Bianca's play does feature a brick-built being that hearkens back to classic LEGO bricks, it also features characters from infantilized Duplo and gendered LEGO Friends products that are differentiated from standard LEGO by design and marketing. Reinforcing this ideological incompatibility, Finn and Bianca have contradictory play styles and aesthetics. The toxic masculinity that Rex brings to Finn's play is inherently rather than inadvertently antithetical to Bianca's attachment pleas. Thus, Finn's rejection of Bianca's play primarily takes the form of coding these two sub-media of LEGO as intrinsically monstrous.

This is perhaps unsurprising, as the strong cultural conditioning surrounding play means that boys are often explicitly encouraged to reject such play. As Kuznets notes of the evolving relationship between boys and dolls:

> The weight of the anecdotal evidence suggests that dolls can be as significant in the early lives of boys as in the early lives of girls, but that dolls' later significance, after a boy's "successful" passage through the straits of Oedipus, is barely recognized. During latency boys may be permitted only an aggressively hostile relationship to the doll, which has become the epitome not only of the dangerous female "other" but of the even more dangerous internal female as well. Children's books about doll play are likely to show boys who will be boys as destructive. (1994, p. 100)

a

Fig. 6.4 A comparison of two representative sets from *The LEGO Movie* and *The LEGO Movie 2* product lines. Whereas most of the sets from the original films depict locations, vehicles, or scenes from the plot (see Lee 2019), approximately two-thirds of the sequel's sets are small, character-centric sets that privilege dramatic play. Seemingly assuming that the centrality of girls' play in the sequel means that these sets should target girl consumers, the two sets have very differently gendered package templates. **a** The sets from the original film are presented in different shades of blue and feature a film strip trailing from the film logo, emphasizing their status as film adaptations. Here, Emmet's Construct-O-Mech (Set #70814) invites players to further dramatize Emmet's heroic battle stance, despite the films actually showing this moment to be largely unnecessary for Emmet's emotional reconciliation

b

Fig. 6.4 **b** The sets from the sequel are presented in Friends-like purples, eliminate the film strip, and instead show close-up shots of the main characters to imply dramatic play. Here, the Systar Party Crew (Set #70848) invites players to further dramatize the adventure-suspending party scene from the sequel by emphasizing the characters' festive re-designs. Against the logic of the films, these product lines reproduce many of the gendered assumptions and market segmentation discussed in Chapter 3

It is therefore particularly meaningful when, in a second rejection of an Oedipal logic, Finn embraces his ability to play with pink hearts and restores Queen Wa'Nabi to her original form as a personified toy-mediated attachment plea. In both films, reconciliation occurs when Emmet rejects the oppositional framing that generates the central conflict and instead synthesizes contradictory play styles. The moral of both stories is therefore that the LEGO ethos of connectivity—as personified by Emmet—transcends the actual messages of its conflicted and oppositional play themes. Allegorizing its own violent tendencies, LEGO transforms itself from merely a site of adventure play to a site of confronting and contesting the self. The symbolic bricolage of LEGO design thus takes on new significance when marshaled for the psychoanalytic bricolage of attachment play.

The narrative frame of the daydreaming child thereby allows these films to portray a mutual reconciliation between toy and character. LEGO toys mediate miscommunicated attachment pleas, allowing the human characters to restore attached familial relationships. At the same time, the powerful emotionality of attachment infuses the medium with affective potential that is portrayed as transforming its oppositional playscripts into invitations for collaborative attachment play. Thus, these films depict attachment being restored in moments of isolation that privilege material mediation and suggest that toy-mediated attachment pleas speak loudest when this medium of "healthy, quiet play" (Baichtal and Meno 2011, p. 14) is front and center. Consequently, the films suggest that LEGO is not merely a passive *medium* that can neutrally transmit any kind of message but is an active *mediator* whose material design is disposed toward strengthening relational bonds.

Beyond how the animated toys-to-life narratives anthropomorphize the medium, the toys themselves offer embedded playscripts that act on players. As Bernstein notes of toys as *scriptive things*, "a thing forces a person into an awareness of the self in material relation to the thing" (p. 73). Toys, that is, are not merely objects that children happen to invest in emotionally—they are designed to invite or elicit certain kinds of emotional investments. While this ideological scripting certainly contributes to the magic of toys, it is also why toys are so ethically fraught. As commodities, any genuine value toys might have to mediate attachment is problematized by the framing of narratives of attachment as advertising messages. This is compounded by the fact that these toy stories reconcile violent and gendered dichotomies that are implicitly scripted by the very medium they position as its reconciliatory force. And they problematically present LEGO as simultaneously passive enough to be a good communication medium and active enough to interject its corporately-authored scripting into such communication. That is, the films suggest not only that LEGO *can be* a medium of attachment, but that LEGO *is* inherently a medium of attachment that actively conditions its players to deploy its passive connective potential in specific ways.

Animating Toys

Although the psychological explanation of the story as imagined play demystifies the toys-to-life drama, the film reestablishes the toys-to-life mystique by animating LEGO characters even within the live-action scenes. This is central to the ideological construction of attachment play since inanimate toys can only be mobilized for attachment while animated toys can themselves mobilize attachment. After being exposed as inanimate toys, Emmet and Wyldstyle, come alive in a new way and even act on their own in the human world. This twist within a twist offers what Bass El-Wakil calls "a little bit of magic" that makes the viewer think "that world is real too" (McKee and El-Wakil 2014, n.p.).

The magic of toys collides with the magic of film in the multifaceted practice of ***animation***, which names both specific filmic media and the more general principles of "the endowing with life and the endowing with motion" (Cholodenko 2014, p. 101). Players endow emotionally resonant toys with life and motion by animating them in dramatic play. Toy stories, especially toys-to-life narratives further dramatize the magic of this animation, actualizing in fiction life and motion that could otherwise only be imagined. Animated toy stories, which are often intertwined with toys in the history of children's media, even further actualize the materiality of life and motion by visually depicting the coming-to-life of the inanimate toy. *Mobility*, for instance, defines both the character toy and animated character as they reinforce each other in Japanese media mix (Steinberg 2012). Drawing on these parallelisms, the culture of children's media often relies on the mutual animating force of toys and film animation as children are invited to play out animating toys within the transmedia context of canonical animating narratives that paratextually script such animating play.

Thus, while calling Emmet the product of Finn's imagination is not incorrect, it is incomplete. Instead, wavering between fantasy and reality, these films breathe life into their animated toys beyond what the child characters could imagine. Consequently, these films speak not only to the animating potential of the imagination but also to the animating potential of LEGO itself, a scriptive thing that mediates its own animation within imaginative attachment play. This attributes deep significance to LEGO that resembles the ***transitional objects*** that psychoanalysts describe as mediating a child's developmental transition from a drive-related fantasy world to a more realistically ordered world.[16] Yet, whereas the significance of the transitional object derives primarily from the psychic investment of the child, the magical touch of animated toys speaks to a fantasy of toys speaking to the child, thereby realizing in fiction the wish that children might have for toys to intercede in their reality.

LEGO is animated within the human world in the pivotal scene before Finn's reconciliation with his father. Finn has been reprimanded and the father has begun to act upon his authority to undo Finn's creative intervention—the apocalypse foretold in the LEGO world. He has taken and placed the Emmet minifigure on his workbench, declaring Emmet to be an "ordinary, regular, generic construction worker." At this moment, Emmet decides to act. Although he encounters great difficulty in moving himself in a reality where he's a toy rather than a hero, Emmet comes to life in fits and starts and manages to throw himself off the workbench, narrowly evading the suspicious glances of the father.[17] Although this is played to comic effect, both Emmet's movement and the father's glances (more on this later) are crucial to the message of the film. Unlike Neo in *The Matrix*, who gains power over the world of illusion because he is 'naturally' a citizen of the real, Emmet is 'naturally' a citizen of the artificial and should have no life or agency in the 'real' world. Animating the inanimate toy, therefore, transforms what reality means.

Like many toys-to-life narratives, the films ask viewers to imagine a world in which toys in fact have the magical potential that emotionally invested children often imagine they do—a power to act as well as be acted upon. Such narratives are fundamentally nostalgic, playing out less a reality in which magic is real and more a sense of childlike wonder capable of finding the magical in reality and especially play. In Finn's case, rather than simply being content to puppet Emmet to play out his own attachment plea, his emotional investment in Emmet reveals a deep-seated longing to be magically saved by his beloved toy. And, in the world of the film, Emmet literally responds.

[16] This concept was initially introduced by early developmental psychologist and psychoanalyst D. W. Winnicott to describe objects like security blankets or teddy bears that, at certain developmental stages, help children bridge the anxiety precipitated by encountering the harsh objectivity of the real world.

[17] In terms of the plot, this is a pointless gesture. He cannot return from the floor to the LEGO world under his own power, but requires the assistance of Finn to do so, which could just as easily have been accomplished without Emmet's self-movement. Clearly, then, its significance is in the animation itself.

While it is tempting to read this scene as the camera eye adopting Finn's perspective and thereby drawing the viewer into his imaginative world, it is clear that Finn is not merely authoring a private fantasy, but hoping for a miracle he cannot bring about himself. This bit of magic is essential to Finn, who needs the toy to possess life to validate investing his attachment plea in it. If Emmet were merely an instrument of Finn's imagination, it would justify his father's instrumental approach to the medium and the fantasy would collapse. Instead, Finn relinquishes authorial power and wordlessly throws Emmet back into the LEGO world to act as his champion, saying "it's up to you now, Emmet."[18] And no matter how much the father's subconscious desires—his latent playful yearnings—might appear to animate Emmet, he has too little knowledge of Finn's story to even subconsciously extend it. And, while it is tempting to read Emmet as animating himself, this reading becomes much less plausible after the film becomes realistic, exposing Emmet as a toy and revealing his narrative as imaginary. While there is something to be gained from each of these perspectives, none of them can satisfactorily account for the magical realism of this scene: the human perspectives cannot account for the magic, and Emmet's cannot account for the realism. I propose, then, a fourth option: that *LEGO* animates Emmet, and does so for *us*, the viewers.

In the moment of Emmet's coming to life, LEGO leverages its animated toy story to advertise itself as a branded transitional object that bridges fantasy and reality. Certainly, consumers are unlikely to expect fully alive, sentient toys when they shop for LEGO products. And yet, there is substance to this mystical promise. Many advertisements depend on a hazy cause-and-effect relationship in which the mere presence of the advertised product precipitates a dramatic transformation in the consumer's life. Similarly, despite restrictions that require advertisements

[18] He does this by wielding a *magic portal* (see Chapter 4 for a discussion of portals in digital play) that consists in a cardboard tube decorated with colorful designs and LEGO elements. As another moment in which an object from the outside world asserts its authority over LEGO, this is a particularly powerful transitional object that makes manifest the slippage between the two realities, presenting a material foundation upon which the imaginative fantasy can become a kind of reality. Significantly, the inscription "magic portal" directly references a 1989 brickfilm by Lindsay Fleay that is one of the earliest known brickfilms. At the time, LEGO was highly resistant to this kind of work and sent Fleay a cease-and-desist request.

to show human hands animating inanimate toys, many toy paratexts capitalize on the longstanding cultural mystique surrounding toys (including a long tradition of toys-to-life narratives like these animated films) to suggest that toys function as active participants in children's play. Rarely presented as a blank slate upon which children can inscribe any meanings they like, LEGO consistently presents itself as a scriptive thing disposed toward cultivating its particularly LEGO ethos of ***development, imagination, creativity***. In this case, these films animate LEGO in such a way that the medium itself almost mystically promises to mediate emotional attachment. As the film wavers between reality and fantasy, it is the LEGO medium—a toy in reality and the world in the fantasy—that binds real and virtual together.

In this context, Finn invests his attachment needs within the LEGO medium, which does more than passively transmit his message. Instead, the medium actively transforms and even performs its own intervention into the attachment relationship. The LEGO toys come alive, animate themselves, and alter circumstances to bring Finn and his father together as if LEGO itself were actively reconciling childhood and adulthood. In a sense, LEGO serves as both passive medium and active ***mediator*** in reconciling father and son. Playing out this dynamic, one can argue that LEGO rather than Finn eventually attunes the father to Finn's attachment needs. As the father is unable to receive Finn's more verbal attachment pleas, one might speculate (as psychologists sometimes do) that his rigidity is grounded in his lack of secure attachment from his childhood. Thus, as Finn is unable to excise his father's childhood demons, only the nostalgic, transitional medium of LEGO can speak into the father's past. In the aforementioned scene, the film juxtaposes Emmet's coming to life with the suspicious glances of the father as he looks over at Emmet, perceptually attuned to a toy magic he consciously denies. As the father hears—albeit distantly[19]—Emmet's stirrings (when even Finn does not), this scene makes tangible the stirrings of the father's heart. Although the

[19] Although the faintness of this perception mostly fits into the convention of toys-to-life narratives such as *Toy Story* in which toys have life only when outside direct observation, that it happens at all begins to undermine this convention. Interestingly, LEGO has produced licensed *Toy Story* products that render the variety of toys mixed in the film into the single medium of LEGO.

father never sees the toy move the way the viewers do, this near-miss radically transforms the father's perspective. The film, that is, reanimates the father's childlike nature by animating the inanimate toy.

Immediately and without any additional prompting, he turns a renewed gaze over Finn's creation (Fig. 6.3) and responds to Finn's attachment plea—a reawakening of attachment desires against which his rigidity was presumably a defense. What presumably stirs for the father in this moment is a sense of nostalgia that reminds the father that toys—as transitional objects—can be magical. Thus, Finn and his father meet by crossing a gap which Adult Fans of LEGO call *the Dark Ages*, the period between putting away one's LEGO as a child and picking it up again as an adult, representing maturation as the loss of wonder. Rather than Finn's attachment pleas restoring the father's relationship to LEGO toys, LEGO itself reminds the father of his latent childlike sense of wonder, restoring his ability to perceive and respond to Finn's attachment pleas. By restoring this wonder, the mystical quality of the LEGO medium creates a space for renewing attachment bonds, another sense in which LEGO actively mediates attachment.

The sequel extends and expands this trajectory by again animating LEGO characters in ways unexplainable from within the framework of the child's daydreaming. In a touch of dark magic, Rex allegorizes the danger of abandonment morphing into toxic masculinity by coming to life out of sheer anger at being forgotten. As Rex explains that "anger was the key to my freedom," the film shows Emmet coming to life. Then, as Rex narrates "I was real," the film shows Emmet actively transforming himself into Rex to literally become a self-made man (Fig. 6.5a). While this narrative certainly allegorizes Finn's growing emotional disconnection, this animated narrative suggests a strangely magical causality in which Finn is not shown driving—either consciously or subconsciously—this transformation. Instead, attributing all the agency to an animated Rex, the film flirts with the notion of darker emotional forces impinging on children from the psychologically-resonant realm of the animated toy.

Counteracting the force of this dark magic, Wyldstyle comes alive to sing a message of hope that captures Finn's attention (Fig. 6.5b), despite her being constitutionally immobile and contextually trapped in the storage bin. In this moment, Emmet comes alive to fight back against Rex. Interestingly, neither Wyldstyle nor Emmet is remotely as successful moving themselves as Rex is—their animating force is more emotional, drawing on the filmic characterization of the characters as heroes of

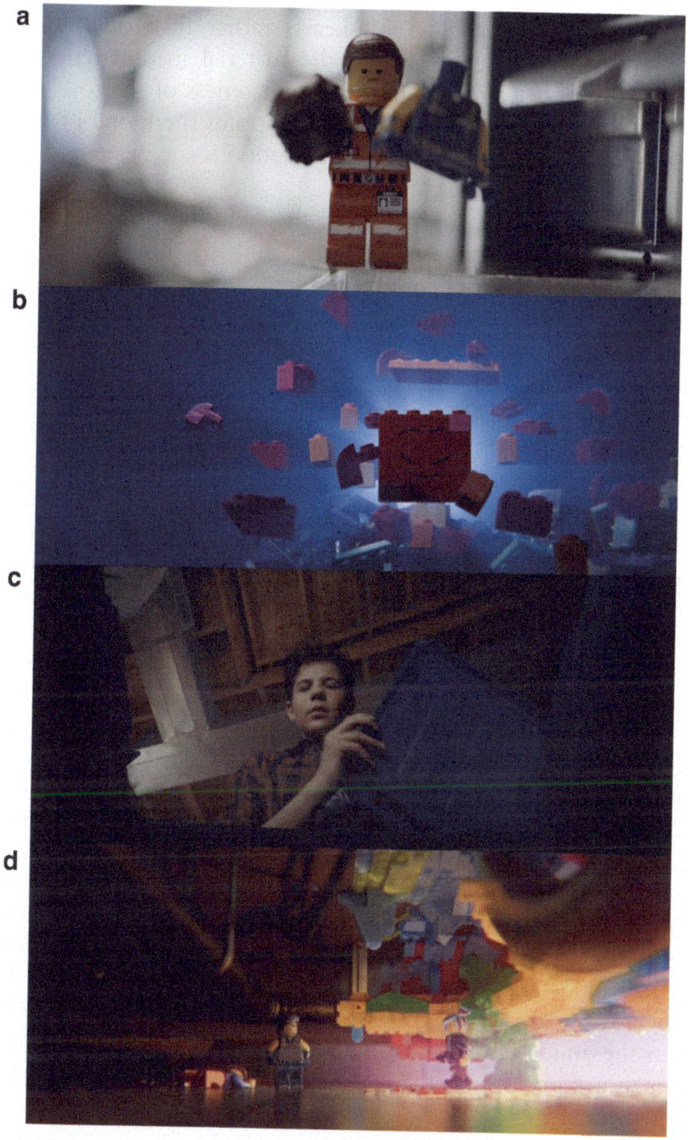

Fig. 6.5 Four screenshots from *The LEGO Movie 2* depicting **a** an embittered Emmet discovering the vest and hairpiece that facilitates his transformation into Rex, **b** Queen Wa'Nabi reassembling herself, **c** Finn opening the storage bin after somehow hearing Wyldstyle's, emotional musical plea, and **d** Wyldstyle approaching Rex as the space under the dryer transforms into LEGO

connectivity who act on children's hearts rather than on the material world. Here, the film acknowledges the material inanimateness of the LEGO toys while stressing the magical animateness of the toys' emotive potential. Thus, the toy character Emmet narrates the moral of the story atop the aforementioned wordless reconciliation scene between Finn and Bianca.

Finally, in the filmic coup-de-grace, the film animates LEGO bricks in a particularly magical way. As Wyldstyle and the other LEGO characters lay trapped and despondent in the storage bin, bricks float up around them and reassemble themselves into Queen Wa'Nabi (Fig. 6.5c), causing the other characters to regain their ability to move. As the personification of the medium itself, Wa'Nabi reanimates not only herself but the entire animated LEGO world. And this reanimated LEGO world subsequently reanimates reality itself. Wyldstyle intercedes in the very moment when Rex has defeated and disassembled Emmet in the space under the dryer that represents the site of his transformative abandonment. After Rex crows his denial of toy animation, saying "This isn't even happening. It's all just the expression of the death of imagination in the subconscious of an adolescent," Wyldstyle proves the transcendence of toy animation. She enters the scene as a figure of animating transformation rather than military reinforcement, stepping beneath the dryer haloed in a transformative light as the real-world space magically transforms into a brightly-colored brick-built environment (Fig. 6.5d). The very site of abandonment becomes reanimated as a site of LEGO play.

The reconciliation in both films, therefore, depends upon the autonomous animation of LEGO, which actively cultivates an ethos of connectivity that reverses how divergent play within the medium causes family members to drift apart. In these toys-to-life narratives, toys not only bear emotional investments, they come alive to actively transform attachment relationships. The toys themselves are a reconciliatory force. While their animation more literally owes to the filmic medium than to any self-moving capacity of the toys, these toy stories are not meditations on the animating force of film. Instead, these toy stories speak to the animating force of inanimate toys, advertising LEGO as a medium of attachment.

The Magic of Toys

A story with wheels within wheels, the LEGO quest narrative turns out to be mediated by the children's imaginative attachment play which turns out to be mediated by the filmic toys-to-life narrative. The final such wheel is that the film is mediated by the logic of the physical LEGO toys, which are the driving force behind the entire film. Materially, the filmic animation mimics the affordances of actual LEGO and weaves actual LEGO products into its world. Ideologically, the films rely on toy-centric central themes like creativity and connectivity to function as a meditation on the medium itself (see Lee 2019). While many toys-to-life narratives invoke commercial products, they often emphasize the inanimateness of the product to instead emphasize the animating force of the filmic or imaginative frames. The LEGO films take the opposite approach, calling into question the animating force of film and imagination to instead locate the deepest layer of agential meaning in the toys themselves.

Underneath all these animating forces lies one originary source of toy magic: Emmet's minifigure body. This physical reality anchors this nostalgic narrative. Alongside intimate shots depicting the father's encounter with his son's creation, the aforementioned scene (Fig. 6.3) contains some of the most intimate shots of Emmet's body, revealing a crucial detail: on his left leg Emmet wears an identification badge that bears his name. This detail shatters the distinction between real and imagined worlds well before Emmet's movement does.[20] When the father insists that Emmet is "just a construction worker" and even when Finn stubbornly maintains that "he's the hero," both ignore the fact that Emmet's identity—like that of the legendary golem—is indelibly etched on his body. Etched, that is, by LEGO, who emerges as the third authorial voice in the attachment play between father and son.

By authoring Emmet to be the hero of Finn's imaginative play, the films reveal how LEGO toys are designed to script attachment play. To position LEGO as a medium of attachment, the films animate real, licensed products that slightly predated the film release and already

[20] In fact, the overlapping of the two realities has been alluded to throughout the film, although it doesn't come full force until the live-action sequence. Examples include the "8 ½ years later" timeframe that points to Finn's age, the quick flashes of "The Man Upstairs" when Emmet first encounters the Piece of Resistance, the role of ordinary objects in the otherwise purely LEGO world, and even the barely visible presence of scratches and fingerprints on the minifigure characters.

contained embedded symbols of attachment. Like toy play, the films are narrative bricolages of already-significant elements built into the toy medium. The films animate toys that, in turn, animate the films. Similar to the twist that reveals the toy story to be Finn's imaginative construction, the films themselves are predicated on a twist that reveals their own narratives to be imaginative extensions of the already extant potential of material LEGO toys. Thus, rather than being a "generic construction worker," Emmet's material reality always already names him as a unique character toy who brings specific narrative potential to attachment play. Emmet has an implied subjectivity that personifies the LEGO ethos of connectivity within its very material form. While an Emmet minifigure cannot actually move itself as his animated counterpart does, these toys are scriptive things that broadcast their preexisting identity to the player.

Unlike some accounts of human subjectivity, Emmet's identity does not flow from depth to surface as if his soul animated his body. Instead, Emmet's body animates his soul in a characteristically LEGO movement from surface to depth. Pointedly, Emmet's name derives from a Hebrew word (פםא or *emet*) meaning "truth" that speaks to the animating influence of this inscription. In Jewish folklore, this word is written on a slip of paper and placed in the mouth of a clay figure to create a *golem*, an animated construct of raw material. Likewise, LEGO brings Emmet to life with a simple inscription—an inscription, moreover, which cannot be attributed to Finn's (or the player's) imagination.

Similarly, the transformation of Emmet into Rex Dangervest in the sequel can only be explained by the intervention of LEGO's corporate authorship.[21] Although Rex claims to have 'found' his new outfit, the outfit he scrounges is visibly designed by LEGO as a play on Emmet's orange safety vest (a "danger vest"). The transformed design of this outfit both negates Emmet's identity by removing the animating identity badge—one 'kills' a golem by erasing the א, transforming *emet* (truth) into *met* (death)—and animates Rex's identity by adding a similarly life-giving inscription of the letter "R."[22] Thus, these minifigure characters are firstly authored by the embedded significations in their physical toy

[21] This corporate authorship is heightened by the persistent dramatic irony of filmic references that would be unfamiliar to Finn, most notably the many references to other film roles played by Chris Pratt, the actor who voices both Emmet and Rex.

[22] Furthering this name-driven transformation, as Rex dons the vest with the significant "R," he re-christens himself as "Rex: Radical Emmet Xtreme."

design, secondly by the animating influence of the films, and only thirdly by the imagination of the playing child Finn. This makes LEGO toy stories quite different from common toys-to-life stories like *The Velveteen Rabbit*, in which the emotional investment of the child is portrayed as providing the sole animating force that fuels the magical transformation.

Thus, when Finn plays out an emotionally-laden narrative as if it were his own creation, he turns out to be following a script already prepared for him, poignantly illustrating the problematic ideology of LEGO's embedded playscripts. While Finn certainly voices his attachment needs through Emmet, he can do so only because LEGO in general and Emmet in particular are already designed to mediate this attachment. After all, Finn's message reaches his father at least in part because the medium already resonates with him, which is not surprising for a medium with a longstanding cultural association with connectivity, family, and play. Similarly, Finn finds a construction worker hero amidst his father's non-fanciful city because LEGO already put it there.

The magic of toys is not that players can speak through them, but that there is a sense in which toys speak *to* players. Toys have a special form of animateness or life, a kind of mediating object-agency. Or, rather, toys have a special form of *after*life, as toy players reanimate toys whose material design and paratextual narratives have already animated them. As attachment play is performed through LEGO bricolage, emotionally-laden LEGO toys are reanimated from a creative reassembly of already-significant elements that point to an underlying ethos of connectivity scripting LEGO play.

Thus, the afterlife of the LEGO toy necessarily reanimates commodity fetishism, reanimating specters of corporately-authored playscripts within the bricolage of attachment play. The films' message of connectivity is at once a profound meditation on how talismanic play objects can mediate connection and an unusually complex toy commercial, that is, a psychic fantasy and a commercial fantasy. Recalling Alan Cholodenko's (2014) theory of *lifedeath* in animation (see Chapter 5), there is a Frankenstein-esque sense in which the reanimated toy passes from life to death to life again as the animating play of the LEGO company is concretized within the commodity and subsequently remobilized by the player. This is the paradox of all scripted LEGO play—that LEGO empowers players agential play by asking players to replicate predetermined scripts as bricolages of predetermined elements.

While the animateness of these toys allow them to deeply engage subterranean emotional needs, the fact that this animation is largely due to corporate authorship calls into question whether attachment play is truly psychic self-expression or is merely a scripted performance like other forms of socializing dramatic play, including the gendered performances (see Chapter 3) that threaten Finn and Bianca's attachment. The LEGO brand of attachment play, that is, problematically situates emotionality within consumerist ideology. Although there are anti-consumerist messages on the surface layer of the film plots, the more the surface layers are peeled back to reveal deeper layers of meaning, the more subtly and powerfully the message of attachment becomes entangled with LEGO-based consumerism.

At the second layer of the toy story narrated by the daydreaming child, the suggestion that the emotional power of the narrative lies in the imagination of the playing child also makes the advertising message more potent by obscuring it. Celebrating the attachment potential of the child makes it seem like agency lies primarily with the consumer. This is an extremely powerful advertising technique, as consumers will often value products that make the consumers feel powerful over products that feel powerful in themselves.

At the third layer of the filmic animation, the film animates LEGO as coming to life to accentuate the mystical potentiality that elevates toys from props to children's companions. Here, LEGO not only symbolizes desirable sentiments but infuses an aura of emotionality in and around its hard, plastic toys. The promise that such toys can easily come to life for children because they are already animated within LEGO's transmedia universe is part of how these films further commodify the medium by construing it as a medium of attachment.

And finally, at the deepest layer, the films reveal that animating potentiality resides first and foremost within the bodies of the plastic toys. Here, all of these layers come together to suggest that these material toys not only passively mediate but also actively promote emotional attachment. The toys script the adventures they are to be used to dramatize, the emotionality they are to be invested with, and the animation they are to be imagined to possess. Indeed, this scripting *is* their animation, the ideological object-agency they exert on players, the subtle invitation to become complicit in the stories the toys are already telling.

Although these successive layers speak less and less directly to capitalism or consumerism, they more and more strongly align the medium with the sentiments of emotional attachment. After all, advertising does not only advertise product features but commonly tries to build emotive associations with products. As Gillian Dyer explains, in advertising "the meaning of one *thing* is transferred to or made interchangeable with another *quality*, whose value attaches itself to the *product*" (qtd. in Gray 2010, p. 28). Yet, while it is quite common for advertisements to draw emotional associations of this kind, the LEGO films make a stronger, more theoretical claim that treats this emotional association as if it were a tangible product feature.

This claim works in two primary ways. First, the material connectivity that defines the LEGO medium—constituting what Bernstein (2009) would call a "determined script," one whose enactment is required to engage the thing as what it is designed to be) is presented as both figuratively and literally mediating emotional attachment. Figuratively, LEGO functions as a material metaphor for attachment. Beyond this, LEGO claims to literally mediate attachment by providing a collaborative playspace where, for instance, father and son can bond by playing together (LEGO has mounted entire advertising campaigns based on father/son play). Second, reinforced by the films and other paratexts, LEGO is advertised as bearing emotive investments of a brand that identifies itself according to an ethos of connectivity. The values behind LEGO design, that is, are advertised as embedded in the commodity. The magic of toys, in a sense, parallels the magic of advertising as both are grounded in the experience of things coming to life to actively contribute value—emotional as well as practical to consumers.

In the end, the primary attachment being constructed by these films is the attachment between consumers and the LEGO brand. As the minifigure and human characters pursue their own emotional attachment bonds, the overall structure of the films cultivates an association between the medium and the concept of attachment itself. LEGO is a medium of attachment only insofar as LEGO is emotionally invested in—toys, after all, typically only come to life when well-loved.

This is not to say that there is no magic in the toy. This narrative of attachment is compelling because LEGO does have potential for attachment play. However, this magic is an ideology that is at least partially constructed by LEGO as a particularly consumerist scriptive thing. The magic of the LEGO toy is, in other words, the magic of a material medium mediating its own ideological construction. The life and afterlife of such toys is therefore a reanimation of the ideological specters that animate the medium as an active force in its own construction. While the reanimating performance ultimately rests with the players, this performance necessarily invokes the emotional investments players have with the medium through a complex form of consumerist attachment born of the emotional pushes and pulls of participatory media culture.

Post-script 6—The Art of Detachment

Although its playscripts suggest that LEGO is a medium of emotional attachment, the underlying materiality of the construction system resists this serendipitous association between material and emotional attachment. At the material level, attachment and detachment are—from certain perspectives—virtually indistinguishable. In many atomistic material encounters, connections are reversible and impermanent. To be an assemblage is to dwell amidst the ever-present possibility of disassembly.

Consider Nathan Sawaya's *Red*,[23] a LEGO sculpture depicting a humanoid figure reaching upwards from within a pool of loose red bricks. As a static sculptural form, *Red* can alternately be conceived as a figure emerging from the component bricks or disintegrating back into them. Here, what distinguishes attachment from detachment is *perspective*. Consequently, this evocative figure can be alternately interpreted as representing birth or death, attachment or detachment. Thus, while the prevailing optimism of the LEGO brand one-sidedly links its material interconnectivity with emotional attachment,[24] on the periphery of LEGO play many notable artworks develop emotional depth by leveraging LEGO to instead explore feelings of loss, anxiety, or emotional detachment. This Post-Script explores several such works that in different ways show that LEGO has more emotional range than official LEGO

[23] *Red* is one of three installments in a triptych that includes Sawaya's most famous work, *Yellow*, which I explore in "The Plastic Art of LEGO" (2014). Although this triptych was initially conceived as representing birth, life, and death, Sawaya notes that *Yellow* "is also one of those pieces whose meaning has changed for me over time" (2015, p. 90). As with *Red* the meaning can be fully inverted, as Sawaya explains "I think now, with the benefit of a decade's hindsight, that *Yellow* is not about death at all, but about opening yourself up to the world without fear or reserve. It's about dropping the mask and the hundreds of little compromises you make every day to show the world the true you, and letting life come at you as it may" (p. 92).

[24] The notable exception is the destructiveness that Rex teaches Emmet. The film, however, clearly situates the LEGO ethos against this kind of deconstruction.

suggests and thereby call into question the inexorability of LEGO playscripts.

While it is possible to outline a relatively coherent ethos of attachment for core LEGO play, sourced in LEGO design and reinforced in LEGO paratexts like the films, it is impossible to map the immense variations that constitute peripheral detachment play. Entire genres of fan LEGO, such as "post-apocalyptic" and "horror," consistently subvert the ethos of attachment. Similarly, contemporary artists occasionally deploy LEGO to symbolize less utopic cultural values.

Notably, Polish artist Zbigniew Libera's provocative[25] LEGO concentration camp sets mimic commercial LEGO set designs to make an artistic statement on the dehumanization and objectification that characterize the Holocaust. Detachment is a central feature of this art, which probes the paradox of the perpetrator—how the bodily intimacy of physical violence is made possible only by extreme emotional detachment from the perpetrator's own embodied acts. As Ernst Van Alphen argues, "Libera deploys a different technique so visitors get a small taste of identification with the perpetrators. His *LEGO Concentration Camp Set* stimulates visitors of the museum or the gallery to envision the possibilities of building their own concentration camp" (2001, p. 75). More than a meditation on the psychology of the perpetrator, this art is an invitation to viewers to psychologize our detachment from historicized atrocities. By decomposing the playful plasticity of LEGO, this art blurs the fine line between anthropomorphizing objects and objectifying human subjects to encourage audiences to reflect on their complicity in treating social systems like playspaces.

A more intimate play of attachment and detachment can be found in *In Pieces*, a mixed-media[26] collaboration that brings together Sawaya's LEGO sculptures with Dean West's hyper-realistic photography. These anachronistic images present striking scenes with key objects strategically

[25] Not only provocative for the viewer, the way these sets mimicked LEGO product design and stated LEGO sponsorship (which was true in a material sense because LEGO provided Libera with the bricks for his art) drew the ire of the LEGO company. LEGO launched a lawsuit against Libera but later dropped the suit presumably due to public response.

[26] Although the primary way LEGO builds circulate is through photography on various media platforms, this project actually mixes the media rather than using one to mediate the other.

replaced with LEGO sculptures, creating an uncanny impression of pixelated LEGO objects that are often more dynamic feeling than the posed human figures.

Bus, for instance, depicts a racialized encounter around public transit reminiscent of the Rosa Parks story. An old-fashioned Black woman and a young, hip blonde woman stand at a lonely bus stop in rigid and uneasy postures as they stand spaced apart and facing opposite directions. Although this art is linked to social issues like Libera's, the picture it paints is much more personal and human. As the symmetry of the background frames this mutual separation, their only point of connection is the young woman's brick-built dog curiously exploring the older woman's shoes, adding a sense of dynamic movement that contrasts with the rigid poses of the human characters. Thus, the doubly inhuman (animal and inorganic) figure offers the only possibility of attachment in this otherwise frigid and depressing scene, perhaps because its existence lies outside the human racial politics that arrest the scene.

A similar interplay of motion and stasis is seen in another piece in this series, *Dress*, which paints a picture of a cold, lonely woman in a vibrant, flowing red dress standing motionless outside a movie theater on a dark and snowy night. As the vibrancy of the dress and brightness of the theater contrast the pathos of the woman, the flowing skirts—almost impossibly crafted from LEGO—illustrate not dynamic motion but the buffeting of the icy wind as they become a plastic prison contributing to the woman's immobility. Materializing both stasis and detachment, the flowing skirts trail off into component bricks, as if the icy wind is dissolving the brightly-colored plastic illusion of the red dress threatening to reveal the underlying naked loneliness of the woman.

While these artworks paint haunting pictures of detachment from the outside, asking viewers to reflect on the nature of detachment itself, several fan works elicit even more personalized forms of self-expression that probe feelings of detachment from the inside. Two particularly expressive pieces are *Unchain My Heart* by Vancouver builder Paul Hetherington and *Worlds inside of me* by Swedish builder Malin Kylinger (a.k.a. Blue Builder). Although these fan builds are as introspective as Libera's art, they reflect less on the detachment of social objectification and more on detachment from sociality.

Unchain My Heart (Fig. 6.6a) draws upon a steampunk aesthetic to create a bust of a strikingly inorganic automaton holding his organic heart in one hand and using the spinning blades the other to sever the chains

a

Fig. 6.6 a *Unchain My Heart* by Paul Hetherington

that hold the heart to his chest. This clockwork figure is also a structure, housing a male minifigure and robot character alike trying to let in a female minifigure who is coming from the surrounding desert toward a walled-off heart. Significantly, this narrative locates emotional attachment in a figurative material detachment as "the heart beats to emancipate itself from its chains to find true love…." (Hetherington 2015, n.p.). While detachment is presented here as freedom, it also represents the death of the inorganic dwelling that is the most expressive figure in this artwork.

b

Fig. 6.6 **b** *Worlds inside of me* by Malin Kylinger (Images courtesy of the artists)

Whereas steampunk art often explores the becoming-inorganic of the human body, here becoming human entails detaching interconnected parts of a cyborg body. Consequently, while this build adopts the LEGO ideal of attachment by celebrating the struggle to communicate attachment needs and find love, it also troubles the uncomplicated picture of attachment in the films by playing with the simultaneous separability and inseparability of humanized inorganic elements and dehumanized organic elements. Thus, this play of attachment and detachment probes the tragic paradox of emotional connection requiring internal disconnection.

Contrasting with Hetherington's inorganic detachment play, Kylinger's *Worlds inside of me* (Fig. 6.6b) is a strikingly organic self-portrait of a woman divided, the two halves of her face set apart by mismatched hair and eyes as she is cloaked with a similarly divided world—one side dark and menacing and the other light and airy. Challenging the trope of an angel on one shoulder and a devil on the other, the two sides organically touch and connect to each other, both within the representation of her physical body and through the worlds of her own creation. Thus, Kylinger describes the meeting of these conflicting emotions as a "pendulum" (2018, n.p.). She even offers another, alternate interpretation: "my experience of having a dual personality where I one moment love all social activities and a busy life and the other moment want to live alone in my tower and just build with my Lego" (n.p.). Whereas LEGO only mediates attachment in the films, here Kylinger reflects on LEGO as also a medium of solitude and detachment. LEGO is at once a way of reaching out to people and a refuge from people.[27] Indeed, perhaps the most pointed part of this reimagination of LEGO's attachment narrative is how she destigmatizes its potential for detachment. In a build that can alternately represent internal division and emotional balance—with the opposing poles rounding out a complex personal identity—Kylinger paints attachment and detachment as an ebb and flow necessary for wholeness.

Without suggesting that detachment is somehow more profound than attachment, what these artworks offer in contrast to the ethos of the LEGO brand is a clear articulation of how the medium offers more than a single message. While the LEGO films are surprisingly nuanced for animated children's films, they script a singular and disturbingly consumerist picture of the attachment potential of this emotionally resonant medium. Similarly, the branded messages explored in this project reveal a surprising dearth of possibilities—not because LEGO is incapable of more but because the ideological work of the LEGO brand is to cultivate certain styles of play over others. Perhaps, then, the art of detachment is as much about dissociating the possibilities of LEGO play from the messages of the LEGO brand as it is about exploring the psychological potential behind conflicted emotional relationships. Deconstructing LEGO is, after all, an exercise in questioning the implicit ideology of a meaning-making system whose meanings are often prepackaged. To creatively reassemble these already-meaningful elements is certainly an art—or, perhaps, an act of play.

[27] Despite the ethos of connectivity described in Chapter 6, LEGO is in many ways more oriented around solitary building. As LEGO set designer Jamie Berard notes, "I think most fans see building as their alone time, and bricks are so valuable to them, either on a sentimental or monetary level. To build in a group setting is a very different dynamic" (qtd. in Bender 2010, p. 182).

Works Cited

Ackermann, Edith, David Gauntlett, Thomas Wolbers, and Cecilia Weckström. 2009. *Defining systematic creativity*. Billund: LEGO Learning Institute.

Bachelard, Gaston. 1994. *The poetics of space*, trans. Maria Jolas. Boston: Beacon Press.

Baichtal, John and Joe Meno. 2011. *The cult of LEGO*. China: No Starch Press.

Bender, Jonathan. 2010. *LEGO: A love story*. Hoboken: Wiley.

Bernstein, Robin. 2009. Dances with things: Material culture and the performance of race. *Social Text* 101 (27): 67–94.

Bowlby, John. 1969. *Attachment and loss: Vol. 1. Loss*. New York: Basic Books.

Bowlby, John. 1973. *Attachment and loss: Vol. 2. Separation anxiety and anger*. New York: Basic Books.

Campbell, Joseph. 2008. *The hero with a thousand faces*, 3rd ed. Novato: New World Library.

Cholodenko, Alan. 2014. "First principles" of animation. In *Animating film theory*, ed. Karen Beckman. Durham: Duke University Press.

Freud, Sigmund. 1995. Creative writers and day-dreaming. In *The Freud reader*, ed. Peter Gay, 436–443. New York: W. W. Norton.

Goggin, Joyce. 2018. How do those Danish bastards sleep at night?: Fan labor and the power of cuteness. *Games and Culture* 13 (7): 747–764.

Grobovaite, Dalia. 2017. Politics of bricolage and the double-sided message of The LEGO Movie. *CJMS* 15: 57–78.

Gray, Jonathan. 2010. *Show sold separately: Promos, spoilers, and other media paratexts*. New York: New York University Press.

Hetherington, Paul. 2015. LEGO steampunk, Unchain my heart. Flickr. https://www.flickr.com/photos/14964802@N07/30592460905/. Accessed 24 April 2020.

Holloway III, John Daniel. 2016. Everything is [not] awesome: Critique and embrace of ideology in *The Lego Movie*. *The International Journal of the Image* 7 (3): 55–63.

Kuznets, Lois Rostow. 1994. *When toys come alive: Narratives of animation, metaphorphosis, and development*. New Haven: Yale University Press.

Kylinger, Malin. 2018. Worlds inside of me. Flickr. https://www.flickr.com/photos/bluebuilder/31079405307/. Accessed 24 April 2020.

Lee, Jonathan Rey. 2014. The plastic art of LEGO: An essay into material culture. In *Design, mediation, and the posthuman*, ed. Dennis M. Weiss, Amy D. Propen, and Colby Emmerson Reid, 95–112. Lanham: Lexington Books.

Lee, Jonathan Rey. 2019. Master building and creative vision in *The LEGO Movie*. In *Cultural studies of LEGO*, ed. Rebecca C. Hains and Sharon R. Mazzarella, 149–173. Cham: Palgrave Macmillan.

The LEGO Movie. 2014. Directed by Phil Lord and Christopher Miller, Warner Bros.

The LEGO Movie 2: The second part [Film]. 2019. Directed by Mike Mitchell, Warner Bros.

Mattes, Ari. 2019. Everything is awesome when you're part of a list: The flattening of distinction in post-ironic LEGO media. In *Cultural studies of LEGO*, ed. Rebecca C. Hains and Sharon R. Mazzarella, 73–95. Cham: Palgrave Macmillan.

McKee, Robert, and Bass El-Wakil. 2014. Has *The LEGO Movie* killed the hero's journey? Wordpress. http://mckeestory.wordpress.com/2014/03/04/has-the-lego-movie-killed-the-heros-journey/. Accessed 24 June 2014.

McLuhan, Marshall. 2005. *Understanding Media*. 2nd ed. Routledge.

Sawaya, Nathan. 2015. *The art of the brick: A life in LEGO*. San Francisco: No Starch Press.

Steinberg, Marc. 2012. *Anime's media mix: Franchising toys and characters in Japan*. Minneapolis: University of Minnesota Press.

Van Alphen, Ernst. 2001. Playing the Holocaust. In *Mirroring evil*, ed. Normal L. Kleeblatt, 65–84. New York: Rutgers University Press.

Varul, Matthias Zick. 2018. The cultural tragedy of production and the expropriation of the brickolariat: *The Lego Movie* as consumer-capitalist myth. *European Journal of Cultural Studies* 21 (6): 724–743.

Varul, Matthias Zick. 2019. The accursed second part: Small-scale discourses of gender and race in *The LEGO Movie 2*. In *Cultural studies of LEGO*, ed. Rebecca C. Hains and Sharon R. Mazzarella, 123–145. Cham: Palgrave Macmillan.

CHAPTER 7

After Words: The LEGO Sandbox

Consistently inconsistent, the defining characteristic of LEGO is that it is mixed and remixed. Mapping this shifting landscape is like taking the topography of a sandbox, a geography of forces and flows rather than sharply defined spaces. As a medium of bricolage, the LEGO sandbox offers play possibilities far beyond its scripted ideological content. Just like a real sandbox, the LEGO system is at once a flexible meaning-making tool and a regulatory playspace—the sandbox, after all, is a technology of physical containment and visual surveillance, as well as a profoundly evocative intersection of tactility and imagination. The paradox of the sandbox is, after all, the very paradox of the LEGO set, the contradictory act of designing a box for thinking outside the box.

To navigate the shifting sands of LEGO play, this project charted five conditions of possibility for meaning-making that govern patterns of movement through LEGO play. As LEGO is a continually expanding and evolving medium with continually proliferating messages, the point of this mapping cannot be to chart a fixed topography that does not exist. Instead, the point can only be to facilitate the ongoing navigation of constantly shifting topographies. More like wilderness training or orienteering than mapping, this payoff is largely methodological—the assumption being that deconstructing the corporate ideology behind LEGO design provides a critical incitement to creative, tactical interventions into the medium.

J. R. Lee, *Deconstructing LEGO*,
https://doi.org/10.1007/978-3-030-53665-7_7

After all, LEGO cannot be reduced to any of the ideologically-laden forms of play—*construction, dramatic, digital, transmedia,* and *attachment play*—discussed in the preceding chapters. Instead, the medium and messages of LEGO should be understood as constituted by the bricolage of perpetually mixing and remixing ideological elements. Thus, these "After Words" reflections suggest that the dominant form of LEGO play is none of the aforementioned forms of play but rather an assemblage of them all (and more). Or, more precisely, that the core of LEGO play is *sandbox play*, a form of meta-play—a bricolage of bricolages—designed to facilitate countless variations and interventions.

Such paradoxical meta-play is possible because remixing is central to both LEGO core and peripheral play. This is the allure of the ideal of the *box*, the personalized playspace where designed sets become dissociated into a brightly-colored flow from which individual creations emerge. Despite constituting the everyday reality of much actual LEGO play, the box is no less evocative for it, as Seth Giddings observes in his survey of reported narratives of LEGO play:

> "The box" is an evocative focus for a multiplicity of memories, and the wellspring from which many LEGO play events emerge, and its collection or absorption of numerous sets, negates critique of themed sets and instructions and constraining. Not only does the box mix up initially distinct sets, it often originates in, or has incorporated, LEGO from older siblings, relatives or secondhand shops, jumble sales, and today, eBay. LEGO, particularly in the amounts needed for fully satisfying play, is not cheap, and old and new pieces and boxes often move through families and communities and down through generations. (2014, pp. 257–258)

Part of the emotional potency of the box is that it represents a personalized material history of localized and potentially intergenerational play. As Jonathan Bender recounts, one of his builds "means something more to me because it consists of bricks from my entire collection" (2010, p. 191). In this way, the LEGO sandbox accrues layers of significance even beyond the extant significances built into individual sets and elements.

Whereas traditional bricolage reclaims the extant significances of scavenged objects, the bricolage of sandbox play is predicated on nostalgic accumulations of significance. As the same elements are mixed and remixed, they operate like a palimpsest in which the ascribing of new meanings may overwrite old meanings while also leaving an emotional trace. Retaining prior significances while promising new ones, this repository is both a historical archive that retains the deconstructed traces of prior play and design and an affective space where history becomes nostalgia, where solid becomes fluid, and where meanings circulate through a flow of plastic potentiality. In this nostalgic accumulation, the LEGO box operates like a sandbox that bears no physical traces of past play but accumulates emotional resonances each time it transforms into a new or familiar imagined world.

LEGO is not designed *for* the sandbox[1]; it is designed to *be* the sandbox. That is, the LEGO system functions as a built environment that organizes rather than merely occupies space. Moreover, it does this in a distinctively self-reflexive way. Whereas the walls of a traditional sandbox circumscribe play from the outside in, LEGO organizes space from the inside out as an inherently systematic medium with predetermined spatial relationships. Thus, whereas a traditional sandbox exhibits a tension between the regulatory domestic space of the box itself and the abstract flow of the sand, the regulatory pull of the LEGO sandbox is built into its particulate bricks, the material traces of ideological designs echoing in the tangible elements themselves—LEGO is sand that can box itself.

At the same time, as an imaginative playspace, LEGO is also a box that looks beyond its walls. Similarly, the nostalgic as well as physical playspace of the sandbox provides an evocative picture of the imaginative and immersive character of child's play. A slice of beach incongruously realized in suburban environments, the sandbox facilitates otherworldly fantasy play even while the sometimes too-present materiality of the sand grounds this fantasy in a tangible reality. As the sand clings to the playing child, it provides a perversely unshakeable particulate reminder of the materiality of play. Although particles of sand are much finer than LEGO bricks, the already-meaningful elements of LEGO bricolage are akin to sand that particulate fluidity constitutes their formative potential. Consequently, the fluid yet systematic LEGO medium comes to materially signify the play it organizes and inspires—it is not only designed to *be* the sandbox; it is designed to *evoke* the fluidity, creativity, and openness of the sandbox.

[1] In some LEGO instruction books, pictorial instructions tell children that LEGO is not to be used outside (indoor play was one of the ten founding LEGO principles) by drawing a red "X" through a picture of bricks sitting on grass.

At the same time, these echoes are amplified by the many representational LEGO paratexts that typically accompany LEGO elements. Building instructions, product packaging, advertisements, and more suggest that the LEGO sandbox is not a space of isolated, spontaneous creativity but rather a space of culturally-constructed creativity. Sandbox play is almost definitionally creative, imaginative, and performative.

Yet, as bricolage constitutes the underlying foundation of such performances, sandbox play is never purely abstract or culturally decontextualized. In the end, the profundity of the sandbox is not its insulation from cultural meaning, but how it encourages playful mixing and remixing of cultural meanings. After all, the regulatory space of the sandbox is not primarily defined by the impermeability of its walls (many other "child-safe" products are far more structurally prison-like), but by the social norms that govern appropriate play. Similarly, ideologically inflected LEGO elements cannot compel any particular mode of play; instead, they invite cultures of play to take up and perform playscripts that resonate through multiple levels of cultural uptake. More than the sum of its messages, the catalog of its products, or even the totality of all LEGO play, LEGO is a principle and a philosophy born of such mixing.

This mixing of messages is characteristic of each of the aforementioned forms of LEGO play. Through *construction play*, the physical act of reassembling already-significant architectural elements draws LEGO play into the tension between playing out embedded socializing messages and creatively reimagining social systems. Through *dramatic play*, signifiers of identity are remixed within performances that both problematize and reinforce gendered binaries. Through *digital play*, material and virtual bricks play off each other in a complex intermingling of digital and analog mediations. Through *transmedia play*, LEGO performs participatory media engagements based on playing with dis-played signifiers that materially reconstruct fictional worlds. And through *attachment play*, toy-mediated storytelling play transmits and transforms profound affective engagements with and within branded toy worlds.

Beyond the mixings implicit in each form of play, however, the sandbox of LEGO bricolage also remixes these forms of play themselves. Remixing construction and dramatic play, LEGO constructions always imply dramatic performances that reciprocally draw upon physical architectures—housing play begets playing house and vice versa. Remixing transmedia and attachment play, story toys provide material signifiers of transmedia worlds designed to be remixed within affective toy stories while circumscribed by narrative paratexts that provide playscripts for LEGO-specific storytelling. And between these dyads, the material worlds of toy play and the virtual worlds of media play become remixed into the complex geography of the digital analog LEGO sandbox as an entry into participatory media culture.

Remixing all these elements into the contradictory yet complementary space of sandbox play renders LEGO a possibility space for connecting diverse play systems, imaginative storylines, and personal relationships into a dynamic, interactive play experience. This is perhaps the true connective potential of the LEGO medium—not just that already-significant elements interconnect, but that the system itself becomes a site for mixing and remixing entire culturally-significant modes of play.

The fluidity of this remix is powerfully illustrated by Olafur Eliasson's *The collectivity project* (Fig. 7.1), a public art installation that invites passersby to build and rebuild an enormous cityscape that reflects their collective imagination of a utopic city, implying that a city can only be utopic if collectively built. As passersby build from a large collection of white bricks reminiscent of the LEGO Architecture Studio set (Set #21050), this project renders LEGO especially particulate and accessible. In other words, just as a sandbox creates an incongruously fluid playspace within the regimented space of suburban architecture (domesticating the natural world being a defining feature of suburban design), Eliasson's installations create fluid playspaces within the regimented space of the urban cityscape (the project usually takes place in open or only partially-enclosed public spaces, asking denizens to reimagine urban spaces in which they are visibly embedded). Consequently, this project emphasizes the city as an assemblage and constructing the cityscape as an act of tactical[2] bricolage.

As in the sandbox, playful participation lends a kind of analog fluidity to the digital LEGO medium. As *The collectivity project*'s description indicates, "the inevitable entropy of the piece begins to soften the hard edges of the designed structures, and mounds of loose pieces gather in the corners between buildings, a beautiful collective creation takes form" (High Line 2015, n. p.). In this softening, the pixelation of digital LEGO

[2] Michel de Certeau (1980/1988) defines *tactics* as how citizens creatively navigate the strategic design of urban landscapes. Eliasson plays off this idea by asking citizens to apply this potentially resistant kind of thinking to this design, blurring the lines between tactics and strategies.

Fig. 7.1 The 2015 High Line installation of *The collectivity project* by Olafur Eliasson. Note the play of regularity and irregularity that emerges from the collective construction process, which encourages creative deviations within repeated structural patterns. Looking closely, one can discover little personalizing touches within the cityscape, such as letters and a heart. Photo by Liz Ligon (Image courtesy of the artist)

mingles with an analog aesthetic as sculptures morph and evolve each time a piece snaps into place. Here, too, construction play mingles with dramatic play as the act of construction becomes staged performance. And even transmedia play mingles with attachment play as the messages circulating through the media environment blend together to constitute a loosely connected multimodal expression of citizens' desire for attachment—to each other and to the urban environment itself. Consequently, this project simultaneously draws on the mixed messages of the LEGO medium, which imagines the city according to a clearly-defined suburban ethos in its Town and City sets, then invites reimagining the city in its Architecture Studio sets, speaking to the democratic potential of mixing messages through collaborative worldbuilding, paradoxically both capitalizing on and subverting prevailing LEGO ideologies.

Such tactical interventions are possible because this system *of* play is always *in* play, at hand, remixable. As a medium of bricolage becomes meaningful through a creative reassembly of already-significant elements, its meaning is always partially determined and always partially in the process of being determined. Actual play in LEGO, in other words, is a much more complex set of significations than the forms of core play described in this project. As meanings are remixed and layered onto this shifting geography, the medium begins to fracture into a multiplicity of personalized significances. At the same time, it is evident that within this complex system of significances, many of the circulating elements are atoms of ideology that are almost genetically encoded to suggest the patterns of their circulation. The ethics of this circulation is by no means simple. Instead, they consist of constantly shifting assemblages forged by the dynamic interaction between the strategic ideologies of core LEGO play and the tactical interventions of actual play performances. As with the sandbox, what all this means is less 'open to interpretation' than 'yet to be performed.' Perhaps, then, deconstructing LEGO aims not only to critique embedded ideologies but to speculate on the possibilities of forging new ones.

In some ways, there is nothing remarkable about this: participatory media culture is defined by such paradoxes as production becomes intermixed with consumption and user interventions are simultaneously invited and commodified. In other ways, the LEGO sandbox is utterly unique—not because there are no other cultural sandboxes, but because this particular sandbox conditions and channels playful performance through its unique brand of media-specificity. For all its mixed messages and ethical ambivalence, LEGO is very much a brand—and what it means to be a brand is to singularly unify otherwise discordant elements into an identifiable and coherent ethos. The LEGO sandbox, in other words, is at once a place where multiple meanings and meaning-making systems are mixed and remixed and the singular symbol of its own playful remix culture.

The ethical response to such a complex yet singular media phenomenon is difficult to articulate—not because there is nothing to say, but because there is so much. The aim of deconstructing LEGO is, therefore, not to articulate any single ethical critique but to help perform ongoing critical interventions into the dynamic spaces of LEGO play. Thus, as a methodological exploration of LEGO play, this project practices deconstruction—not only to name and thereby loosen the ideological hold of LEGO's embedded playscripts but also to expand the critical resources available for creative bricolage in and around LEGO play. As many artists and builders have shown, the already-significant elements of LEGO play can be creatively repurposed and reassembled in countless constructive ways. In so doing, they reshape the sandbox in which they work. In this spirit, this project has deconstructed several problematics of LEGO play to playfully construct methodological foundations for *critical bricolage*, a creative, transformative, ethical intervention that recontextualizes the already-significant ideological elements of participatory media culture to hopefully reshape that culture.

It is difficult to say more here—not only because there is so much to say, but also because genuinely transformative practices must transcend articulation. After all, media means—forms, performs, transforms—more than it says. Consequently, any formative, performative, or transformative ethical response must also mean more than it says. To leave open a philosophy meant to be assimilated and subsequently kicked away, Ludwig Wittgenstein concludes his *Tractatus Logico Philosophicus* by suggesting that what comes after words is a simple negation: "whereof one cannot speak, thereof one must remain silent" (1922, §7). But what emerges within such silences? What else might come after words? Whereof one cannot speak, touch, build, play.

Works Cited

Bender, Jonathan. 2010. *LEGO: A love story*. Hoboken: Wiley.

de Certeau, Michel. 1988. *The practice of everyday life*, trans. Steven Rendall. Berkeley: University of California Press.

Eliasson, Olafur. 2005. *The collectivity project* [Interactive installation]. Exhibited at Tirana, Albania.

Giddings, Seth. 2014. Bright bricks, dark play: On the impossibility of studying LEGO. In *LEGO studies*, ed. Mark J.P. Wolf, 241–267. New York: Routledge.

High Line. 2015. Olafur Eliasson: The collectivity project. https://www.thehighline.org/art/projects/olafureliasson/. Accessed 24 April 2020.

Wittgenstein, Ludwig. 1922. *Tractatus logico-philosophicus*, trans. F. P. Ramsey. New York: Harcourt, Brace & Company.

Notes on Terminology

Following are some clarificatory notes on the use and derivation of several terms that recur throughout this project. These notes are not intended to function as universal definitions, as terminological usage is highly context-dependent. Instead, these notes clarify how terms are used in this project and to what extent these usages maintain or adapt standard usages.

This project uses LEGO terminology consistent with common usages in LEGO corporate and fan communities. Terminology for naming categories of toys and play is less standardized, so this project names several reasonably common-sense categories (like "story toys" or "transmedia play") in ways that may be phrased slightly differently in other scholarship. Finally, this project borrows quite a bit of critical terminology from various scholarly discourses, so these notes clarify whether critical terminology is used in a specific disciplinary sense or a more general sense (which I often have to do when adapting concepts not initially formulated for discussing toys).

List of Terms

Actual Play—In this project, *actual play* names moments of play that are actually performed by real consumers. In other discourses, *actual play* is often used more specifically to refer to recorded or streamed gameplay experiences, especially for storytelling games like RPGs (role-playing games), produced by game media channels on sites like YouTube and

J. R. Lee, *Deconstructing LEGO*,
https://doi.org/10.1007/978-3-030-53665-7

Twitch. Despite the name, the actuality of these moments of *actual play* can be highly influenced by the context of the streamed performance (a similar argument can be made about the reality of "reality television"). Thus, in this project, I depart from this more specific definition and use *actual play* to name any unrecorded or unobserved moments of play conducted by countless players across the world.

When *deconstructing* the *medium* and *messages* of LEGO, it is vital to remember that the broader *ideological* patterns that run throughout LEGO design do not overdetermine all performances of *actual play*. In *actual play*, individual players exert their creativity to create new play performances that sometimes adhere to and sometimes deviate from the LEGO *playscripts*. As this project explores these *ideological* constructions and *playscripts*, it makes very few claims about patterns in *actual play*. Thus, this project leaves the vast multiplicity of these *actual play* performances to future sociological, ethnographic, or psychological studies. Although such research is beyond the scope of this project, *deconstructing* the cultural positioning of LEGO should complement research on how real consumers respond to this cultural positioning in *actual play*.

AFOL—Acronym for "Adult Fan of LEGO." This term can be used loosely to describe any adult who exhibits fannish behavior (say, an avid collector) or more narrowly to refer to adults who actively participate in the fan community (attending conventions, posting their creations online, etc.). This project typically uses the latter, narrower definition, allowing *AFOL* to function more like a proper name as opposed to the generic category of *fan*.

Analog—As I use it, the adjective *analog* names an informational paradigm that conceives of something according to a "continuous scale" (Cramer 2019, p. 695), such as the continuously sweeping second hand of a watch or the rising and falling mercury in a thermometer. I also sometimes use *analog* as a noun (see *digital analog* below) to indicate a thing that is analogous to another thing.

Animateness—As I use it, *animateness* refers to a sense of life, mobility, and potentiality that even technically inanimate objects can possess. This idea resonates with theories of filmic animation as bringing objects/art to life (see Cholodenko 2014) and *toys-to-life* narrative.

Assemblage—Although *assemblage* is a theoretical term that could easily relate to this project, I use the term in its more ordinary sense to refer to things that are assembled. Since the primary theoretical framework for this project is *bricolage* rather than *assemblage*, I find it useful to maintain a more neutral term for describing material assembly. Theories of *assemblage* often blur the distinctions between subject and object, instead emphasizing the intertwined agency of human and nonhuman "actants" (as they are known in Actor-Network Theory). While future study might productively trace how theories of *assemblage* can describe object relations between LEGO and players, I avoid such reflections to make this project less theoretical and more applied. The more general language of *assemblage* used in this product simply helps clarify its narrower focus on exploring more *media-specific* forms of *object-agency*.

Atomism—In "The Plastic Art of LEGO," I argue that LEGO is an *atomistic* medium insofar as its material construction system mimics the properties of classical Democritean *atomism*, especially how its *indivisible elements* are pieced together in a *vacuum* to produce a *reductive* and *mechanistic* system (2014, pp. 99–100). In that paper, I further draw upon the interplay of *atomism* and *plasticity* in LEGO to pursue a much more philosophical project of performing uniquely material conceptual explorations of materiality itself. In this *deconstructive* project, I use *atomism* and *plasticity* primarily to describe the materiality of LEGO as a medium of *bricolage*.

Attachment—Derived from the psychological *attachment theory* pioneered by John Bowlby and Mary Ainsworth, *attachment* names secure emotional relationships. However, this concept shifts as I appropriate and adapt *attachment theory* to fictional toy contexts. *Attachment needs* name a universal, everyday need for attached relations. *Attachment pleas* are attempts to communicate *attachment needs* to others, not necessarily using verbal language.

Attachment Play—As I use it, *attachment play* names the playful activity of using toys to express a need for emotional *attachment*. This neologism is inspired by psychological research into *attachment theory*, which explores how children need to establish particular kinds of healthy emotional relationships with at least one established caregiver. However, I generalize the notion of *attachment play* beyond central *attachment* relationships recognized in psychological discourse because play may attachment patterns, even in relationships that do not meet the criteria for full-fledged attachment relationships. For instance, it is useful to have vocabulary for how a child at a park may use toys to invite another child into a provisional emotional relationship, even without an expectation that this relationship will meet core emotional needs. *Attachment play* is one of the five main forms of play discussed in this project.

Block—An abstract geometrical solid rendered as a small material object. *Building blocks* name the traditional toy *blocks* that for centuries have been a popular medium of stacking play. *Blocks* differ from *bricks*, which are explicitly architectural. Although LEGO toys are sometimes colloquially referred to as *blocks*, I maintain a distinction between abstract *blocks* and architectural LEGO *bricks* (see Preface).

Brick—A rectangular unit of construction material typically used to construct walls. *Bricks* differ from *blocks* in being explicitly architectural. Some of the earliest and most iconic LEGO elements are rectangular *bricks*, which are sometimes named based on their number of studs (a 2x2 brick vs. a 2x4 brick). These elements were originally named as the Danish equivalent to "bricks" and are still commonly referred to as "bricks" in English.

Brickfilm—Stop-motion LEGO films, usually produced by fans or independent creators. Also known as *brick flicks*. *The LEGO Movie* is not a *brickfilm* because it does not use genuine stop-motion animation, although it does have a similar visual style. *Brickfilms* are a sub-genre of fan video or fanvids, which name any fan-created films.

Bricolage—Derived from a French professional practice of "tinkering" that has gained traction in several scholarly discourses, I define *bricolage* as the *creative reassembly of already-significant elements*. Most actual instances of *bricolage* operate on two levels: a literal level—the *material bricolage* of piecing together physical elements (see *construction play* above)—and a metaphorical level—the *symbolic bricolage* of piecing together images or ideas to create a new expression. A *bricoleur* is someone who practices *bricolage* (I use *bricoleur* to refer to all genders although this term is gendered in the original French).

Canon and Canonicity—In fandom terminology, *canon* names the authoritative texts and narrative details of a story. The *canon* usually consists of officially licensed narratives, but not all official narratives have equal *canonicity*. For instance, the *Star Wars* feature films are considered most *canonical* whereas many of the plotlines of its videogames are considered less *canonical*. *Canonicity* may also change over time as when the *Star Wars* Expanded Universe texts were officially excluded from the *Star Wars* canon when Episode VII was released.

Character Toy—A toy that explicitly represents a (typically fictional) character, often but not always associated with a popular *media franchise*. Not all anthropomorphic toys are *character toys*. For instance, generic dolls and stuffed animals might represent generic girls and dogs instead of specific characters like Barbie and Snoopy.

Computational—As I use it, *computational* names processing that involves electronic computer technology. While in ordinary language, *computational* can also name things like pen-and-paper mathematics, I typically use this term in a more restricted, technologized sense to contrast with *digital*, which I use in a pointedly more general sense to argue that *digitality* need not be technological.

Connectivity—Although I use this word in the everyday sense, I often play with its dual meanings to draw out how material *connectivity* can serve as a metaphor for social *connectivity*. LEGO is literally a medium of material *connectivity*, but also maintains a strong *ethos of connectivity*—how the brand invites and values collaborative play.

Construction Play—As I use it, *construction play* names the playful activity of using *construction toys* to build something. This differs from construction-themed play: a toy bulldozer might thematically imply a construction site without facilitating any actual construction on the part of the player. As a *construction toy*, all LEGO sets facilitate *construction play*, even those not themed after construction sites. *Construction play* is one of the five main forms of play discussed in this project.

Construction Toy—A toy that encourages building or *construction play*. This term is typically used to refer to more complex, systematic toys like Meccano, Erector, K'Nex, Tinker Toys, and LEGO, although it could also include generic building *blocks*.

Creativity—Although *creativity* is a common cultural value, it is not always very clearly defined. In her seminal work *The Creative Mind*, Margaret Boden defines creativity as "the ability to come up with ideas or artefacts that are *new, surprising and valuable*" (1991/2004, p. 1). Boden further notes that these terms are context-dependent—what is new, surprising, and valuable to one person may not be to another. In particular, Boden indexes a key difference in scope between *historically* *creative* ideas that are new to history and *psychologically creative* ideas that are new to specific individuals.

I exclusively use this latter usage of *psychological creativity*, which is much more appropriate for a study of toys. After all, any individual act of LEGO play may not be new (because others may have produced the same construction), surprising (because LEGO can facilitate very formulaic constructions), or valuable (because play is a very humble and local activity) to society at large. To account for the *creativity* of all manner of children's play, this project uses very loose criteria for *creativity*, describing even minimally creative activities as creative. This also fits with the general *phenomenological* orientation of this project, which is concerned with the experience and activity of play.

I also tend to use *creativity* when creative imagination manifests in actual, tangible creation. Whereas Boden defines creativity as applying to both "ideas and artefacts," this study generally emphasizes the latter. This is not because *imagination* (pertaining to ideas) and *creativity* (pertaining to material things) are truly distinct—indeed, the LEGO value *development, imagination, creativity* very clearly links them. Instead, tying the usage of *creativity* more strictly to material creation highlights how even conceptual creativity often has the structure of tangible creation, as in when representational elements are remixed within *symbolic bricolage*.

Deconstruction—Although this term derives from the critical tradition of *deconstruction* established by Jacques Derrida, I use it more generally to name any analytical approach that unravels how its objects of analysis are *ideologically* constructed. To deconstruct LEGO, then, is to analyze it not as a set of neutral objects for making meaning, but instead as a set of objects that already suggest a great deal about what how can and should become meaningful. As the title suggests, *deconstruction* is a central methodology and aim of this project.

"Development, Imagination, Creativity"—One of the *ten founding principles* for LEGO toys penned by second-generation heir to the LEGO Company *Godtfred* Kirk *Christiansen*. In the Preface, I argue that this principle most directly and consistently describes the LEGO ethos across its many products and eras. Notably, *development* speaks to a tradition of treating toy play as intrinsically educational while *imagination* and *creativity* speak to closely related ideals of children as playfully exploring beyond the boundaries of the everyday. So, it is particularly telling that LEGO links these ideals to imply that the more mundane cognitive or educational benefits of LEGO are closely connected to its more evocative and playful aspects.

Digital—As I use it, *digital* names an informational paradigm used to conceive of something as discrete or countable (Cramer 2019, p. 694), such as the numerical display on a watch or thermometer. This is a deliberate departure from the use of *digital* to name computer technology. As I argue in Chapter 4, the common usage of *digital* to mean *computational* is not particularly relevant to a *phenomenological* reading of media, which explores the mediated experience itself rather than the technological tools used to create that experience. Consequently, I argue that *digitality* is fundamental to the LEGO experience.

Digital Analog—I coin this seemingly contradictory term to name how material (analog) systems can manifest (analogize) *digital* paradigms. In particular, I argue that LEGO is a *digital analog* because its hands-on material play operates according to a fundamentally *digital* logic in which discrete and countable LEGO elements are snapped into an implicitly geometrical structure. In general, I believe it is important to recognize how many aspects of modern material culture function as *digital analogs*, showing *digitality* to be a widespread informational paradigm that transcends the computational.

Digital Play—As I use it, *digital play* names the playful activity of using toys to probe the boundaries between materiality and virtuality or analog and digital. I deliberately do not restrict *digital play* to a discussion of computer-mediated play (i.e. videogames) because I believe it is important to name how systematic material toys like LEGO are oriented around a fundamentally *digital* logic. *Digital play* is one of the five main forms of play discussed in this project.

Digitization—The process of rendering *material* objects in virtual *environments*. For instance, one can scan a photograph taken with a film camera to produce a *virtual* image readable by a computer. In this project, I use this primarily to refer to how certain videogames *digitize* the LEGO medium by allowing users to construct with LEGO in *virtual* environments. This also includes the stop-motion LEGO aesthetic in *The LEGO Movie* (see Chapter 6) but does not include computer-animated graphics that have a LEGO aesthetic but do not have an underlying construction system (as in the LEGO *Star Wars* animations described in Chapter 5).

Dis-Play—As I use it, *dis-play* is a play on how *display* (staging objects for visual consumption instead of tangible interaction) contributes to *dis-play* (the negation or denial of the play experience). Yet, although this term frames *dis-play* as against *play*, I argue that they are mutually-constitutive. Many toys are designed so that visually consuming them implies the activity of *play*. Conversely, many toys infuse their *play* with embedded *symbolic* and *ideological* content that their material form places on *dis-play*. So, I do not use *dis-play* as a pejorative term but instead to acknowledge how *play* relies on it the shadow complement of its own negation. I develop and apply this concept primarily in Chapter 5.

Dramatic Play—As I use it, *dramatic play* names the playful activity of using toys (usually but not always *character toys*) to dramatize or perform playful narratives. Although Chapter 3 argues that dramatic play is feminized within the ideological construction of LEGO play, the general concept need not be. In fact, stereotypically masculinized action figures often invite dramatic play. *Dramatic play* is one of the five main forms of play discussed in this project and is also the foundation of *transmedia play*.

Element—Any single LEGO piece or part. Although LEGO *bricks* are the most famous LEGO *element*, there are thousands of LEGO *elements* that are much less brick-like, so I often use this more general term instead. Every LEGO *element* is given a unique Item number by LEGO so it can easily be cataloged, a practice referenced by *The LEGO Movie* in illustrating its vision of creative vision (Lee 2019).

Fan—The definition of a *fan* is often contested both within fan communities and amongst fan scholars. *Fans* can be defined by the strength of their emotional investments in the objects of their fandom, by their participation in fan communities, or by their participation in fan practices. In this project, I use *fan* to loosely name any or all of these things. I use the more specific term *AFOL* (Adult Fan of LEGO) to refer to people who self-identify as members of this particular fan community and actively participate in at least some of its main fan practices (building *MOCs*, posting in online environments, attending conventions, etc.).

Although LEGO is certainly a legitimate fandom, it can look quite different from some other media fandoms. Because LEGO is a *medium*, the most prevalent and publicized forms of LEGO fandom involve creatively using LEGO to build things. Using other media to express fannish investment—such as writing fanfic, performing cosplay, or drawing fan art of *The LEGO Movie*—certainly happens but is not the most recognized form of LEGO fandom. As this project is more focused on *deconstructing* the *medium* and *messages* of LEGO, it leaves a more thorough fan studies approach for future scholarship.

Figure—As I use it, a *figure* is the most general term for a toy that represents a human or human-like entity. Thus, I would say that LEGO *minifigures*, LEGO *mini-dolls*, and all other dolls or action figures are all *figures*. Using *figure* to refer to all these kinds of toys keeps the larger category more gender-neutral, whereas many of these more specific terms are heavily *gendered*. Colloquially, *figure* sometimes also names collectible models rather than toys, but I do not use the term in this way.

Gender—In contrast to biological *sex*, a distinction famously theorized by Judith Butler (1990/2008), *gender* names *performative* cultural ideas and ideals of masculinity and femininity. One might say that *gender* is an *ideological* construct layered onto biological sex. While it is beyond the scope of this project to deeply probe the relationship between sex and *gender*, this project focuses almost exclusively on *gender*, since ideas of biology more than biology itself that inform gendered representation in inanimate toy design.

Ideology—A set of culturally-significant ideas or ideals according to which members of a society may think and act. Many of the *ideologies* I explore in this project are not well-articulated philosophies (like the platform of a political party) but are implicit or assumed (like the widespread but unspoken idea that we should encourage children's creative spirit).

Instructions—Most LEGO *sets* include printed *instructions*, which can be variously called "building instructions," "instruction manuals," "instruction booklets," and so on. These *instructions* outline a step-by-step process of building the LEGO *set* using numbered pictorial representations of the *set* in various stages of construction. A typical Step 1 would simply show a LEGO baseplate or bottom-most *element*. Then, the next step would show that element with a few pieces snapped into place and so on, so that a visual comparison between any two adjacent steps would provide an implicit direction for how to build that step.

Int(e)ractability—In "The Plastic Art of LEGO" (2014), I coin this term to name a paradox at the heart of materiality—that objects are *interact-able* precisely because they are *intractable*. Even for an ordinary natural object like a rock, the intractable properties that make it resist human interaction (inertia, weight, hardness, etc.) are precisely the properties that make it able to be interacted with. For a designed, systematic object like LEGO, the intractable constraints built into the construction system (such as the rigid, gridlike placement of the studs) are precisely what provide its constructive affordances—that is, its ability to be interacted with.

In "The Plastic Art of LEGO," I argue that having a toy exhibit this degree of *int(e)ractability* provides unique possibilities for performing hands-on material philosophy. In this project, I mostly consider the *int(e)ractability* of LEGO as how it materially mediates its ideological play—how, for instance, LEGO simultaneously gives its players fixed *messages* built into the material LEGO designs and invites players to create new interactive meanings from those *messages*.

Interactive Media—Although consumers typically have to do something to interact with any medium (turning the pages in a book, pressing play on a video player, etc.), *interactive media* typically names media that demand a significant degree of user input, such as interactive fiction, toys, board games, and videogames.

LEGO—A flexible term that can refer alternately to the LEGO company (officially named The LEGO Group), the LEGO medium, LEGO bricks, and so on. I capitalize LEGO throughout this project not because it is the preferred typography of the company, but as a reminder that LEGO is a fundamentally corporate medium. As it is only "LEGO" that names the brand and company, I do not extend this practice to LEGO product lines that are also presented by the company in all capital letters. Thus, I use "Technic" instead of "TECHNIC," "Mindstorms" instead of "MINDSTORMS," and "Bionicle" instead of "BIONICLE."

LEGO Foundation—A nonprofit foundation funded by but run independently from the LEGO company. Its mission is to promote creative development and learning through play—sometimes but not necessarily related to LEGO play. Throughout this book, the *LEGO Foundation* primarily appears as the sponsor and publisher of research reports that summarize and propose theories of play which I argue speak to the ethos behind LEGO toys. As the *LEGO Foundation* now includes the former LEGO Learning Institute, I refer only to the *LEGO Foundation* even when a report was initially published under the LEGO Learning Institute.

LEGO Friends—Released in 2012, *LEGO Friends* is a series of LEGO products explicitly targeted at young girls. The line features a group of five core Friends who live in Heartlake City. While *Friends* is a play theme like any other (City, Ninjago, etc.), it also introduced a new aesthetic and new Mini-dolls and therefore has become a partially distinct medium from traditional LEGO. To underline how *Friends* functions both as a specific product line and as a spinoff LEGO sub-medium, I sometimes also refer to the entire range of 'for girls' subthemes (Elves, DC Super Hero Girls, Disney Princess, etc.) as within the umbrella of the larger *Friends* sub-medium.

LEGO Principles—In 1963, Godtfred Kirk Christiansen penned *ten principles* to govern LEGO products. The full list (Baichtal and Meno 2011, p. 14) is as follows:

- Unlimited play potential
- For girls and for boys
- Fun for every age
- Year-round play
- Healthy, quiet play
- Long hours of play
- Development, imagination, creativity
- The more LEGO, the greater the value
- Extra sets available
- Quality in every detail.

Licensed Toy—A toy designed and marketed according to a licensing agreement between different corporate entities. Thus, LEGO *Star Wars* is a licensed toy because LEGO had to negotiate the rights to make these toys with the holders of the *Star Wars* intellectual property. LEGO Ninjago is a *media toy* but not a *licensed toy* because the media property it relates to is also a LEGO production, so no licensing agreement is necessary.

Material/Materiality—I use *material* primarily in the everyday sense of consisting of tangible matter. *Materiality* names the quality of this existence, which includes material *int(e)ractability*. Sometimes, I also use *material* to contrast with its natural complement: *virtual*.

Materiality is a concept theorized across many interrelated disciplines, including Material Culture studies, Thing Theory, Object-Oriented

Ontologies, Actor-Network Theory, and so on. Furthermore, many disciplines exhibit a "material turn" when they direct renewed attention to the materiality of their objects of study. This project is inspired by several of these discourses but tries to maintain the broad everyday use of the term without aligning itself with one particular definition from these fields.

Media Culture—A culture that has grown around particular ways of using media. For instance, while television could theoretically function as a medium in any number of cultures, particular places and times have distinctive *media cultures* that relate to television in specific, historically-situated ways. The *media culture* of television includes all the practices, rituals, and discourses surrounding how individuals and societies interact with television.

Media Ecology—A term that has evolved and been used variously throughout the history of Media Studies, *media ecology* most broadly names a shift in emphasis from thinking of media just as transmission technologies to thinking of media as complex environments. Theories of *media ecology* typically assume the interconnectedness of culture, media, and technology, and I use the term to refer to this general insight rather than any one specific theory.

One way this term can vary is in its scope of application. In the seminal media theory of Marshall McLuhan (1964/2005), such ecological thinking was typically tied to the cultural impact of a particular medium. For instance, one might think of *media ecology* as naming how, for instance, the rise of home television sets precipitated shifts in a cultural media landscape. In more contemporary contexts, *media ecology* can also name the ecological relationships that media form with each other.

In this project, I apply this term more generally and variously across different contexts and scopes. For instance, I use *media ecology* to sometimes refer to larger cultural trends such as *participatory media culture* and sometimes to refer to specific *transmedia franchises* such as LEGO *Star Wars*. In these instances, I use the term less as a strictly-defined category and more to emphasize the ecological nature of media environments.

Media Franchise—Strictly speaking, *media franchise* (see Johnson 2013) refers to the business practice of *franchising*—where a larger franchiser like McDonalds provides smaller franchisees the rights to conduct business using their broader umbrella typically in exchange for a fee or share of the profits—applied to media companies. As this project deconstructs the medium and messages of LEGO toys rather than the formation of the LEGO company, I typically use *media franchise* is a looser, more colloquial sense to name both a collection of linked media texts and the corporate forces (licensing agreements, branding, etc.) that link them together. For instance, I would describe *Star Wars* as an example of a *media franchise* because there are many different *Star Wars* media texts and products that are all marketed under the single *Star Wars* brand. *Star Wars* and LEGO *Star Wars* are both also *transmedia franchises* because they span multiple media.

Media-Specific Analysis—N. Katherine Hayles (2004) proposes *media-specific analysis* as a methodology of analyzing texts that accounts for the specific materiality, affordances, and constraints of the medium. For example, instead of analyzing a novel only in terms of the story, Hayles compares how readers interact with physical and electronic books to see how the unique material form of these books shapes the reading experience. This is a particularly important methodology for studying LEGO, a medium where materially engaging with the medium is at least as central to the play experience as the content. While Hayles argues that material experience is significant for all media, the cultures surrounding media like books often tend to privilege content over materiality.

Media-specificity names the specific material affordances and constraints that define a medium. For instance, the *media-specificity* of LEGO includes how the elements fit together, their visual appearance, and so on.

Mediateness—As I use it, *mediateness* refers to the distinctive ways a *medium* transmits and transforms its messages into an experiential phenomenon. The *mediateness* of LEGO encompasses the material, conceptual, and ideological qualities of how it can be used to make meaning.

Media Toy—As I use it, a *media toy* is any toy situated within a larger *media franchise* with significant investment in non-toy media. For instance, the Kenner *Star Wars* action figures and LEGO *Star Wars* toys both derive much of their significance from the popular film franchise. Many *media toys* are also *character toys* and *licensed toys*, although the LEGO *media toy* expands beyond both of these categories as its own media franchise based in *construction play*.

Medium/Media—A *medium* is a *material* (and sometimes *virtual*) system through which meanings can be recorded, transmitted, and/or expressed. *Mass media* such as books, film, and television are typically able to disseminate content for wide one-to-many circulation. *Artistic media* such as painting and sculpture typically have a single original that is designed to be experienced in close physical proximity (although *artistic media* are often reproduced in other wider reaching media). I argue that LEGO is a toy medium because it uses its toys as a system for recording and transmitting meanings. LEGO is more like an *artistic medium* than a *mass medium* in that it is very hands-on and is not well equipped for transmitting meanings, so it often relies on *mass media* like photography or film or *social media* like Flickr or YouTube to be shared widely. LEGO is also a *popular medium* in that it is designed to circulate within everyday consumer culture.

Message—What a *text* communicates or expresses through its *medium*. On the surface level, this includes the most explicit content related in a *medium*, such as the plot of a movie, the information in a newspaper, or the *theme* of a LEGO set. On a deeper level, the *message* can also include implicit *ideological* promptings like how certain *play themes* are targeted toward differently *gendered* players. Finally, as media theorist Marshall McLuhan (1964/2005) famously argued, *the medium is the message* in that media networks are culturally impactful beyond any specific information they transmit. I argue that for LEGO, *the medium is the message* in a particularly strong sense, as one of the things that LEGO expresses is the characteristics of its *medium*. LEGO, in other words, consistently weaves together these three types of *message*—*messages* about its *thematic* content, its *ideological* promptings, and its *mediateness*.

Miniature—More than a synonym for "small" or "diminutive," I use this term to allude to a centuries-old Western tradition of *miniatures*, collectible replicas primarily marketed for adult display. Closely related to the history of the dollhouse, early *miniature* houses or "cabinet houses" were not designed for children's play but as decorative displays of craftsmanship for wealthy families. Yet, as this tradition strongly overlaps with the dollhouse tradition, *miniature toys* name toys that in some way resemble this tradition of *miniature* display. For instance, the detailed models used in contemporary *miniatures* games certainly owe an inheritance to this tradition although they are not explored in this project. As a *construction toy*, the LEGO System differs from traditional *miniatures*, but many of its sets do reflect a *miniature* aesthetic.

Most importantly, many LEGO sets play with the *miniature sublime*, which is how I describe the distinctive allure of the miniature tradition. The *miniature sublime* is a pleasurable and/or profound space of contradiction derived from the simultaneous experience of being invited into the fantasy world of the *miniature* and being excluded from the material scale of the *miniature*. As with the Kantian sublime, the *miniature sublime* relies on a paradoxical interplay that emerges when the immense scope of human conception and perception confronts the material smallness of its own embodied standpoint. This is playfully embodied in *miniature* objects, which typically derive pleasure from making this profound philosophical contradiction immediately and viscerally present to everyday experience.

Mini-doll—A redesigned LEGO figure released in 2012 that replaces the *minifigure* in the 'for girls' LEGO Friends product lines. As this name suggests, these figures were designed and marketed to more closely adhere to the feminized doll tradition than previous LEGO minifigures. *Mini-dolls* and *minifigures* are compared in Chapter 3.

Minifigure—A brick-built LEGO character with rotatable legs, arms, wrists, and head. LEGO *minifigures* were initially released in 1978, replacing earlier designs of buildable LEGO figures which in turn replaced non-buildable toy figures in earlier sets.

MOC—Acronym for "My Own Creation," which is what some members of the *AFOL* community call their fan builds.

Object-Agency—As I use it, *object-agency* names the distinctive way material objects exert influence on their users and environment. Some theories of material culture, such as Actor-Network Theory, typically use language that blurs the lines between the agency exerted by human subjects and the agency exerted by material objects, both of which are sometimes called "actants." While I appreciate this theoretical point, in the interests of clarity I use the term *object-agency* to mark the activity of nonhuman "actants."

Paratext—Coined by Gérard Genette and extended by Jonathan Gray, *paratext* refers to any of the supporting materials or texts surrounding a particular text. When analyzing LEGO toys as a primary text, its main *paratexts* are building instructions, product packaging, advertisements, web content, magazines, and other LEGO media (film, television, videogames). However, which text is supporting which can be a matter of perspective—one could easily consider the material LEGO sets as *paratexts* to the feature films.

In this project, I typically treat material LEGO as the primary text and all other LEGO media as *paratexts*. I maintain this emphasis because it is the materiality of LEGO that is most undertheorized within the study of LEGO and most distinctive in comparison to other Media Studies analyses. So, I believe it is particularly important to emphasize material LEGO toys within this emerging scholarship, even though it is certainly the case that non-toy LEGO media are culturally impactful and merit their own textual analyses.

Participatory Media Culture—Popularized by Henry Jenkins, *participatory media culture* is the name I most commonly use to describe the 21st-century media culture I am writing from (although I write from the U.S. American context, *participatory media culture* is a more global phenomenon). This term—which is also often phrased simply as *participatory culture*—names a trend of increased ability for consumers and users to produce their own content that then circulates within the larger *media culture*. This includes both DIY versions of traditional media (like fanfic as homebrew novels or fanvids as homebrew movies) and social media conversations around more traditional media. *Participatory media culture* does not specifically name an increased emphasis on *interactive media*, but I suggest that *interactive media* like LEGO can shape media participation in significant ways.

Performativity—More generally, social theories of *performativity* describe how performing certain actions or behaviors can also *ideologically* construct identity (these theories are inspired by a philosophy of linguistic *performativity* that this project largely does not address). The most common specific application of this idea is *gender performativity*, a term popularized by feminist theorist Judith Butler (1990/2008) to describe how social patterns of performing *gender* construct the concept of *gender*. When a biological female acts in a recognizably feminine way, for instance, she is *performing* a culturally-constructed ideal of what women in a certain culture are supposed to be; without that cultural context, her actions would be essentially meaningless. I adapt this theory in a somewhat unusual way by applying it to the design of toy bodies, which falls outside the scope of most theoretical discourses on *performativity*.

Phenomenological—This project often adopts a *phenomenological* approach that describes a given event in terms of how it is experienced through human perception and conception. This approach loosely generalizes the philosophical school of phenomenology typified by the work of Martin Heidegger and Edmund Husserl and applies it primarily to media *phenomena*.

Plasticity—As I use it, LEGO exhibits *plasticity* in several ways that go beyond its *plastic* substance. First, LEGO reflects an ideology of modern *plasticity*, a post-industrial tendency to view the world as a reserve of resources that humans can mine and reconfigure at will. Second, the LEGO construction system reflects a sculptural *plasticity*, a field of continuous or fluid possibilities. Despite being materially *atomistic*, particulate LEGO elements come together in such a large array of combinatory possibilities that the medium feels malleable, moldable, and sculptural. In a sense, the *plasticity* of LEGO is the systematic binding of its *atomistic* elements, the interconnective potential that allows these already-significant elements to be creatively reassembled through *bricolage*.

In "The Plastic Art of LEGO" (2014), I draw upon the interplay of *atomism* and *plasticity* in LEGO to pursue a much more philosophical project of performing uniquely material conceptual explorations of materiality itself. In this *deconstructive* project, I use *atomism* and *plasticity* primarily to describe the materiality of LEGO as a medium of *bricolage*.

Play—Although it is one of the most central terms in this project, I deliberately do not define *play* as distinctly as some other critical terms. I do this firstly to maintain the multifaceted nature of its everyday usage, which I believe is a good reflection of the way that LEGO facilitates varied forms of *play*. Any specificity concerning *play*, therefore, is indicated by compounds like the five main modes of *play—construction play*, *dramatic play*, *digital play*, *transmedia play*, *attachment play*—explored in each central chapter.

Playset—A physical structure or frame designed and marketed for staging toy play. LEGO *sets* are commonly designed for constructing *playsets* to be used for *dramatic play*. These *sets* typically include both the elements for building the *playset* itself and the figures for staging the drama. For some other toys, *playsets* are sold separately from *figures*.

Procedural Rhetoric—Analogous to verbal rhetoric, *procedural rhetoric* names how *interactive media* can make 'arguments' by guiding participants through meaningful procedures in which they discover new things about the possibility space of a system that often reflects systemic real-world issues. This term was coined by Ian Bogost (2007) and has become a well-known framework for videogame studies. Although Bogost does not emphasize toys like LEGO or paratexts like LEGO instructions, the concept of *procedural rhetoric* easily applies in these contexts. For instance, LEGO building instructions provide a procedure to follow that not only guides builders through reproducing a set but also offers implicit messages about LEGO and LEGO construction.

Product Line—A series of sets designed and marketed as thematically linked. For example, the LEGO *Star Wars product line* groups together all *Star Wars* sets as a smaller category within the full range of LEGO products. These sets all bear not only the LEGO brand marker but the brand name of the product line. Because sets in a single *product line* are designed to resemble each other, many of the analyses in this project can be reasonably generalized across a *product line*. For instance, I make arguments about LEGO *Star Wars* vehicles that reasonably apply to the full range of vehicles, which typically share remarkably similar design features. I sometimes abbreviate *product line* simply as *line*, as in the "LEGO *Star Wars* line."

Remediation—From a book by Jay Bolter and David Grusin (1999) of the same name, *remediation* names a process by which new media reconfigure (rather than merely replace) prior media forms. As I am less concerned with the broader history of media technologies remediating each other, I typically use this term in a more localized sense. For instance, if I say that LEGO toys *remediate* a film, I mean that they express not only the content of the film (plot, character, setting, etc.) but also some of its filmic qualities, such as when a LEGO set mimics the architectural staging of a film set.

Script/Playscript—Drawn from Robin Bernstein's (2009 and 2011) theory of a *scriptive thing*, a *script* is how the material design of a thing communicates implicit instructions to interact with it in particular ways. As Bernstein (2009) metaphorically explains, a *script* is how a thing *invites* *us to dance*. For example, the material design of a book invites readers to read it in a particular way (flipping pages and reading text from left-to-right and top-to-bottom in English). Bernstein calls this kind of *script* a *determined script* because this kind of script must be followed for the thing to function as designed. A thing might also have more *implicit scripts* that invite participants into certain additional relationships not required for the thing to function. *Implicit scripts* are more varied and can include complex cultural and ideological resonances.

Most modern toys have strongly implied *playscripts* that suggest the proper or intended way to play with a toy (for instance, baby dolls are designed in various ways to elicit a desire to care for them). I often refer to playscripts as *embedded* because they are contained in the material design of the toys themselves. One could argue that many embedded toy *playscripts* mix *determined* and *implicit scripts* because they simultaneously reveal what makes a toy a toy and present additional implied *ideological* messages. However, *playscripts* can also be reinforced by *paratexts* like advertising and product packaging that might directly depict children playing with the toy in particular ways.

Set—Most LEGO products are sold in *sets*, boxed groupings of elements with building instructions that show how the elements are to be put together. Most LEGO *sets* are also *playsets*. Each *set* has a unique identifying number.

Story Toy—As I use it, *story toy* refers to any toy designed to facilitate storytelling. While most *character toys* and *media toys* are implicitly also *story toys*, I often use this term to emphasize when toys more explicitly encourage storytelling (in contrast to how a *character toy* or *media toy* might, for instance, be primarily marketed as a collectible).

Storyworld—All stories have settings, places in which the plots take place. The particular settings narrated in a story, however, also imply a larger world known as a *storyworld*. This *storyworld* is an imagined or *virtual* world that transcends any particular story set in that world. Famous storyworlds include Oz, Wonderland, Middle Earth, and the *Star Wars* universe, all of which imply vast spaces full of interesting things that extend well beyond their original stories (encouraging sequels and spinoffs). These examples are more fantastical, but even a historical novel has a *storyworld*, albeit one that closely resembles our own.

Stud—The cylindrical protrusions on the top of LEGO bricks. As part of the *stud-and-tube* system, studs snap into the empty spaces between and inside the cylindrical tubes to hold bricks together. The word "LEGO" is embossed on every LEGO *stud*.

Symbolic—I often use the phrase *symbolic bricolage* to refer to the *creative reassembly of already-significant meaning-making elements*. As I use the terms, *symbolic* elements are concrete particles of meaning that are plausibly interpretable by people who share certain cultural literacies. Although *symbolic* elements can include linguistic elements like words, I more commonly refer to *icons*—conventionally established signifiers like stop signs that visually communicate clear, culturally-recognized meanings—as well as visual representations that operate based on resemblance. LEGO design operates in both these ways, printing visual icons on its bricks and designing the shape of bricks to represent real-world objects.

Text—Although the word *text* has many different meanings in everyday usage, the one that I use most commonly is that of a media *text*, a single work in a medium. If film, print, and photography are media, a single film, book, or photograph would constitute a media *text*. This is a term of convenience that works in many circumstances although it can be quite difficult to draw coherent boundaries around a *text*. For instance, a film like *Star Wars Episode IV* might seem to constitute a single *text*, but that single film is accessible in a wide variety of nonidentical formats. The film viewed in widescreen is subtly different from the film viewed in 4:3; the film viewed in the theater is a quite different experience than the film viewed on a streaming platform.

Consequently, grouping multiple media products under the heading of a single *text* is useful only so long as the variations between these products do not substantively alter the analytical claims made about the text. For this project, films and LEGO sets are relatively stable *texts* that exhibit only minor variation. It is worth noting, however, that videogame *texts* (especially *LEGO Worlds*, which has undergone substantive updates since its Steam prerelease) are more likely to vary in ways that potentially disrupt their analysis.

Theme—A *theme* names a general organizing framework for play based on fictional or real-world topics. For LEGO, a single *product line* will typically have several related but more specific play *themes*. For instance, the LEGO City *product line* includes *themes* such as firefighting, construction, and NASA. *LEGO Friends* includes *themes* such as restaurant work, pet care, and leisure activities.

Town Plan—The original organizing principle for LEGO toys was the *Town Plan* (1958–1966), which featured a variety of enclosed suburban structures, a playmat with streets and plots to organize play, and non-buildable toys such as cars and traffic cops. The *Town Plan* is analyzed in Chapter 2.

Toy—In keeping with common usage, I use *toy* to name material objects of play that are typically (but not exclusively) marketed toward children. The category of *toys* generally excludes games, which structure material play through a *virtual* set of rules, although this distinction is not absolute. For instance, some LEGO Ninjago *toys* stage contests of battling tops that have some gamelike rules and goals.

Toy Medium—As I use it, a *toy medium* is any toy system that has the meaning-making capacity to function as a medium unto itself. Many construction toys are also *toy media*, but few have as strong a case as LEGO, which can be used to build a variety of structures that leave quite different impressions. It is harder, for instance, to construct something that is not a log cabin out of Lincoln Logs. This is not to say that Lincoln Logs cannot be analyzed as *toy media*, but rather that LEGO offers a wider range of creative possibilities more analogous to non-toy media.

Toys-to-Life—A genre of fiction about physical toys coming to life, whether through magic or imagination. I rely on Lois Kuznets' (1994) *When Toys Come Alive* for background on literary toys-to-life narratives, although this project does not directly focus on this genre. Instead, this project more commonly references how this narrative tradition has extended to other media such as film and videogames, where *toys-to-life* has come to refer to a specific genre of videogames (including *LEGO Dimensions*), in which physical toys are integrated into the game interface (see Chapter 4).

Toy Story—As I use it, *toy story* refers to stories about toy play. *Toy stories* have a slightly different emphasis from *toys-to-life* stories: the former typically emphasizes the player, while the latter typically emphasizes the toys. Ironically, this means the film *Toy Story* is more of a *toys-to-life* story than a *toy story*.

Transmedia—*Transmedia* is a widely-used but variable descriptor that has generated several related terms (see below). In its most general sense, *transmedia* describes content developed to circulate across multiple media networks, especially when this content synergizes well (like when spinoff novels extend a film plot). The way contemporary media industries are trending, most popular *media franchises* have some *transmedia* component, so this term can quickly lose specificity. However, there are some advantages to this generality. I use a broader, looser definition of *transmedia* to account for the fact that media toys, which have often been excluded from *transmedia* scholarship, are often clearly linked with *transmedia* texts in contemporary cultural practices.

Despite the usefulness of this broad definition, it is worth remembering that this term can be contested in several ways. First, as Scott (2010) argues, the typical definition privileges media content created by large media corporations without critically reflecting on how these corporate media networks problematically circumscribe fandom. Second, *transmedia* can only name the intersection of multiple media if those media are considered clearly distinct; however, many media are themselves *transmedial* in complex ways. Third, many media franchises lauded as *transmedial* may develop only weak synergy between its multiple media texts and might perhaps be better described as *cross-medial*. Lastly, *transmedia* may bear a Western cultural bias, as it does not always account for similar phenomena in other parts of the world, such as the Japanese concept of *media mix* (see Steinberg 2012). These and other critiques mean that the contestedness of *transmedia* should be considered part of its central definition—*transmedia* may aim at synergy but is never without tension, whether between its multiple media texts or its multiple definitional valences.

Transmedia Play—As I use it, *transmedia play* names the playful activity of using toys (usually, but not always, *licensed toys*) to dramatize or perform playful narratives that directly engage with familiar media narratives. Drawing on Media Studies discourses around *transmedia*, *transmedia play* is a subcategory of *dramatic play* and is one of the five main forms of play discussed in this project.

Transmedia Storytelling/Franchise/World—As defined by Henry Jenkins (2006), *transmedia storytelling* is a media practice in which multiple stories set in the same universe are told in multiple media. A *transmedia franchise* (see also *media franchise* above) names both this set of interrelated stories and the media company that authors it. A *transmedia world* or *storyworld* is a fictional universe narrated across multiple media. Consequently, any storyworld has the potential to become a transmedia world. In this project, I apply these terms to *media toys* like LEGO although Media Studies historically does not consider toys to be equal participants in these kinds of transmedia engagements, typically reserving these terms for mass media like literature, film, and television. In this, I follow a precedent set by Mark J.P. Wolf, who subtitled the seminal 2014 *LEGO Studies* anthology "Examining the Building Blocks of a Transmedial Phenomenon."

Sandbox Play—As I use it, *sandbox play* names the playful activity of weaving together multiple forms of toy play, potentially including mixing different types and brands of toys to do so. This metaplay is the sixth form of play discussed in this project and the primary focus of After Words.

Virtual—The word *virtual* is often associated with technology, especially computational technologies that create *virtual worlds*, sites of interaction that lack a physical location. I often use the term in this sense, for instance, in comparing *material* LEGO (physical toys) with *virtual* LEGO (toys rendered in videogames). It is important to note that *virtuality* is not exclusive to computational environments. While physical LEGO toys are not exactly *virtual*, the implied Cartesian grid that organizes material LEGO construction is *virtual*—it exists as a structural regularity extrapolated from the material design of LEGO toys but lacks material existence itself. In this broader definition, *virtual* can refer to objects or systems that have no direct material existence, although they may be implied by or interacted with via material systems (as when a material videogame interface allows players to perform actions within an otherwise immaterial *virtual* world).

References

Ackermann, Edith, David Gauntlett, Thomas Wolbers, and Cecilia Weckström. 2009. *Defining systematic creativity*. Billund: LEGO Learning Institute.

Ackermann, Edith, David Gauntlett, and Cecilia Weckström. 2009. *Defining systematic creativity in the digital realm*. Billund: LEGO Learning Institute.

Aldred, Jessica. 2014. (Un)blocking the transmedial character: Digital abstraction as franchise strategy in Traveller's Tales' LEGO games. In *LEGO studies*, ed. Mark J.P. Wolf, 105–117. New York: Routledge.

Alphin, Tom. 2015. *The LEGO® architect*. China: No Starch Press.

Alsdorf, David, and Brian E. Gravel. 2018. Representational Praxes. In *Making engineering playful in schools*, ed. Brian E. Gravel, Marina Bers, Chris Rogers, and Ethan Danahy. Billund: The LEGO Foundation.

Althusser, Louis. 1971. *Lenin and philosophy and other essays*, trans. Ben Brewster. New York: Monthly Review Press.

Anderson, Benedict. 2000. *Imagined communities: Reflections on the origin and spread of nationalism*. New York: Verso.

Armstrong, Frances. 1996. The dollhouse as ludic space, 1690–1920. *Children's Literature* 24: 23–54.

Augé, Marc. 2008. *Non-places: An introduction to supermodernity*, trans. John Howe. New York: Verso.

Bacharach, Sondra, and Roy T. Cook. 2017. Introduction: Play well, philosophize well! In *LEGO and philosophy*, ed. Roy T. Cook and Sondra Bacharach, 1–3. Hoboken: Wiley Blackwell.

Bachelard, Gaston. 1994. *The poetics of space*, trans. Maria Jolas. Boston: Beacon Press.

Baichtal, John, and Joe Meno. 2011. *The cult of LEGO*. China: No Starch Press.

J. R. Lee, *Deconstructing LEGO*,
https://doi.org/10.1007/978-3-030-53665-7

Baker, Neal. 2014. Middle-earth and LEGO (re)creation. In *LEGO studies*, ed. Mark J.P. Wolf, 40–54. New York: Routledge.
Bak, Meredith A. 2016. Building blocks of the imagination: Children, creativity, and the limits of *Disney Infinity*. *The Velvet Light Trap* 78: 53–64.
Barthes, Roland. 1972. *Mythologies*, trans. Annette Lavers. New York: Hill and Wang.
Bartneck, Cristoph, Qi Min Ser, Elea Moltchanova, James Smities, and Erin Harrington. 2016. *PLoS ONE* 11 (5): 1–27.
Beckman, Karen. 2003. *Vanishing women: Magic, film, and feminism*. Durham: Duke University Press.
Bem, Sandra Lipsitz. 1993. *The lenses of gender: Transforming the debate on sexual inequality*. New Haven: Yale University Press.
Bender, Jonathan. 2010. *LEGO: A love story*. Hoboken: Wiley.
Bernstein, Robin. 2009. Dances with things: Material culture and the performance of race. *Social Text* 101 (27): 67–94.
Bernstein, Robin. 2011. *Racial innocence: Performing childhood and race from slavery to civil rights*. New York: New York University Press.
Black, Rebecca W., Bill Tomlinson, and Ksenia Korobkova. 2016. Play and identity in gendered LEGO franchises. *International Journal of Play* 5 (1): 64–76.
Boden, Margaret A. 2004. *The creative mind: Myths and mechanisms*, 2nd ed. London: Routledge.
Bogost, Ian. 2006. *Unit operations: An approach to videogame criticism*. Cambridge: MIT Press.
Bogost, Ian. 2007. *Persuasive games: The expressive power of videogames*. Cambridge: The MIT Press.
Bök, Christian. 2012. Ten Maps of Sardonic Wit. *MIT Unbound Symposium*. https://futurebook.mit.edu/2012/05/ten-maps-of-sardonic-wit-christian-bok/. Accessed 9 April 2020.
Bolter, Jay David, and Richard Grusin. 1999. *Remediation: Understanding new media*. Cambridge: The MIT Press.
Bowlby, John. 1969. *Attachment and loss: Vol. 1. Loss*. New York: Basic Books.
Bowlby, John. 1973. *Attachment and loss: Vol. 2. Separation anxiety and anger*. New York: Basic Books.
Brewer, John. 1980. Childhood revisited: The genesis of the modern toy. *History Today* 30: 32–39.
Buerkle, Robert. 2014. Playset nostalgia: *LEGO Star Wars: The Video Game* and the transgenerational appeal of the LEGO video game franchise. In *LEGO studies*, ed. Mark J.P. Wolf, 118–152. New York: Routledge.
Burden, Michael, and Sean Gouglas. 2012. The algorithmic experience: Portal as art. *Game Studies* 12 (2): n.p.

Butler, Judith. 2008. *Gender trouble: Feminism and the subversion of identity*. New York: Routledge.

Campbell, Joseph. 2008. *The hero with a thousand faces*, 3rd ed. Novato: New World Library.

Cholodenko, Alan. 2014. "First principles" of animation. In *Animating film theory*, ed. Karen Beckman. Durham: Duke University Press.

Cook, Roy T. 2017. Ninjas, Kobe Bryant, and yellow plastic: The LEGO® minifigure and race. In *LEGO and philosophy*, ed. Roy T. Cook and Sondra Bacharach, 91–101. Hoboken: Wiley Blackwell.

Coupland, Douglas. 1995. *Microserfs*. New York: ReganBooks.

Cramer, Florian. 2019. What is "post-digital." In *New media, old media*, ed. Wendy Hui Kyong Chun and Anna Watkins Fisher, 689–702. New York: Routledge.

Cross, Gary. 1997. *Kids' stuff: Toys and the changing world of American childhood*. Cambridge: Harvard University Press.

Cross, Gary. 2010. Valves of adult desire: The regulation and incitement of children's consumption. In *Childhood and consumer culture*, ed. David Buckingham and Vebjørg Tingstad, 17–30. Great Britain: Palgrave Macmillan.

Day, Lori. 2014. The little girl from the 1981 LEGO ad is all grown up, and she's got something to say. *Women You Should Know*. https://womenyoushouldknow.net/little-girl-1981-lego-ad-grown-shes-got-something-say/. Accessed 1 April 2020.

De Certeau, Michel. 1988. *The practice of everyday life*, trans. Steven Rendall. Berkeley: University of California Press.

Deleuze, Gilles. 1992. Postscript on the societies of control. *October* 59: 3–7.

Derrida, Jacques. 1998. *Monolingualism of the other; or, the prosthesis of origin*, trans. Patrick Mensah. Stanford: Stanford University Press.

Dewar, Gwen. 2018. The benefits of toy blocks: The science of construction play. *Parenting Science*. https://www.parentingscience.com/toy-blocks.html. Accessed 17 January 2020.

"digital, n. and adj." *OED Online*, Oxford University Press, March 2020. www.oed.com/view/Entry/52611. Accessed 16 April 2020.

Doyle, Mike. 2011a. The making of... part 1. *Snap*. http://mikedoylesnap.blogspot.com/2011/09/making-of-part-1.html. Accessed on 25 June 2018.

Doyle, Mike. 2011b. Victorian on mud heap. Flickr. https://www.flickr.com/photos/7931559@N08/6192835432/. Accessed 15 April 2020.

Dresner, Eli, and Ofra Rechter. 2016. From symbol to 'symbol', to abstract symbol: Response to Copeland and Shagrir on Turing-machine realism versus Turning-machine purism. *Minds & Machines* 26: 253–257.

Driscoll, Catherine. 2015. The doll-machine: Dolls, modernism, experience. In *Dolls studies*, ed. Miriam Forman-Brunell and Jennifer Dawn Whitney, 185–206. New York: Peter Lang.

duCille, Ann. 1996. *Skin trade*. Cambridge: Harvard University Press.
Edelman, Lee. 2004. *No future: Queer theory and the death drive*. Durham: Duke University Press.
Einwächter, Sophie Gwendolyn, and Felix M. Simon. 2017. How digital remix and fan culture helped the Lego comeback. *Transformative Works and Cultures* 25: n.p.
Eliasson, Olafur. 2005. *The collectivity project* [Interactive installation]. Exhibited at Tirana, Albania.
Faison, Donald. 2010. BlackStormTrooper. YouTube. https://www.youtube.com/watch?v=YePE2tloDd4. Accessed 24 April 2020.
Fanning, Colin. 2018. Building kids: LEGO and the commodification of creativity. In *Childhood by design*, ed. Megan Brandow-Faller, 89–105. New York: Bloomsbury Visual Arts.
FlapJack Films. 2016. Lego Star Wars Special. YouTube. https://www.youtube.com/watch?v=csVqeJodPSk. Accessed 24 April 2020.
Fleming, Dan. 1996. *Powerplay: Toys as popular culture*. Manchester: Manchester University Press.
Formanek-Brunell, Miriam. 1993. *Made to play house: Dolls and the commercialization of American girlhood, 1839–1930*. New Haven: Yale University Press.
Foucault, Michel. 1995. *Discipline and punish: The birth of the prison*, trans. Alan Sheridan. New York: Vintage Books.
Freeman, Elizabeth. 2010. *Time binds: Queer temporalities, queer histories*. Durham: Duke University Press.
Freud, Sigmund. 1989. *Beyond the pleasure principle*. New York: W. W. Norton.
Freud, Sigmund. 1995. Creative writers and day-dreaming. In *The Freud reader*, ed. Peter Gay, 436–443. New York: W. W. Norton.
Garlen, Jennifer C. 2014. Block party: A look at adult fans of LEGO. In *Fan CULTure*, ed. Kristin M. Barton and Jonathan Malcolm, 119–130. Jefferson: McFarland & Company.
Gauntlett, David, Edith Ackermann, David Whitebread, Thomas Wolbers, and Cecilia Weckstrom. 2010. *The future of play: Defining the role and value of play in the 21st century*. Billund: LEGO Learning Institute.
Genette, Gérard. 1997. *Paratexts: Thresholds of interpretation*, trans. Jane E. Lewin. Cambridge: Cambridge University Press.
Geraghty, Lincoln. 2014. *Cult collectors: Nostalgia, fandom, and collecting popular culture*. London: Routledge.
Giddings, Seth. 2014. Bright bricks, dark play: On the impossibility of studying LEGO. In *LEGO studies*, ed. Mark J.P. Wolf, 241–267. New York: Routledge.
Gitelman, Lisa. 2016. "Raw data" is an oxymoron. In *New media, old media*, ed. Wendy Hui Kyong Chun and Anna Watkins Fisher, 167–176. New York: Routledge.

Goggin, Joyce. 2018. How do those Danish bastards sleep at night?: Fan labor and the power of cuteness. *Games and Culture* 13 (7): 747–764.

Goldberg, Daniel, and Linus Larsson. 2013. *Minecraft: The unlikely tale of Markus "Notch" Persson and the game that changed everything*, trans. Jennifer Hawkins. New York: Seven Stories Press.

Gray, Jonathan. 2010. *Show sold separately: Promos, spoilers, and other media paratexts*. New York: New York University Press.

Gregg, Peter B. 2018. Brick-olage and the LEGO/brand axis. *The Popular Culture Studies Journal* 6: 29–44.

Grobovaite, Dalia. 2017. Politics of bricolage and the double-sided message of *the LEGO movie*. *Canadian Journal of Media Studies* 15: 57–78.

Grosz, Elizabeth. 2001. *Architecture from the outside: Essays on virtual and real space*. Cambridge: The MIT Press.

Hains, Rebecca C., and Jennifer W. Shewmaker. 2019. "I just don't really, like, connect to it": How girls negotiate LEGO's gender-marketed toys. In *Cultural studies of LEGO*, ed. Rebecca C. Hains and Sharon R. Mazzarella, 247–269. Cham: Palgrave Macmillan.

Halberstam, Jack. 2018. *Trans*: A quick and quirky account of gender variability*. Oakland: University of California Press.

Hardt, Michael. 1999. Affective labor. *Boundary* 26 (2): 89–100.

Harel, Idit, and Seymour Papert (eds.). 1991. *Constructionism*. Norwood: Ablex Publishing Corporation.

Hayles, N.Katherine. 2004. Print is flat, code is deep: The importance of media-specific analysis. *Poetics Today* 25: 67–90.

Hebdige, Dick. 1988. *Subculture: The meaning of style*. New York: Routledge.

Heidegger, Martin. 1962. *Being and time*, trans. John Macquarrie and Edward Robinson. New York: Harper Perennial.

Hetherington, Paul. 2015. LEGO steampunk, Unchain my heart. Flickr. https://www.flickr.com/photos/14964802@N07/30592460905/. Accessed 24 April 2020.

High Line. 2015. Olafur Eliasson: The collectivity project. https://www.thehighline.org/art/projects/olafureliasson/. Accessed 24 April 2020.

Hills, Matt. 2016. LEGO Dimensions meets Doctor Who: Transbranding and new dimensions of transmedia storytelling? *Icono* 14: 8–29.

Hinck, Ashley. 2019. *Politics for the love of fandom: Fan-based citizenship in a digital world*. Baton Rouge: Louisiana State University Press.

Holloway III, John Daniel. 2016. Everything is [not] awesome: Critique and embrace of ideology in *The Lego Movie*. *The International Journal of the Image* 7 (3): 55–63.

Huizinga, Johan. 1950. *Homo ludens: A study in the play element in culture*. Boston: Beacon Press.

Hunter, Madeleine. 2018. Bric[k]olage: Adaptation as play in *The Lego Movie* (2014). *Adaptation* 11 (3): 273–281.
Hutcheon, Linda. 2013. *A theory of adaptation*, 2nd ed. New York: Routledge.
Jay, Nancy. 1981. Gender and dichotomy. *Feminist Studies* 7 (1): 38–56.
Jenkins, Henry. 2006. *Convergence culture: Where old and new media collide*. New York: New York University Press.
Jenkins, Henry. 2013. *Textual poachers: Television fans and participatory culture*. Updated Twentieth Anniversary Edition. New York: Routledge.
Johansson, Barbro. 2010. Subjectivities of the child consumer: Beings and becomings. In *Childhood and consumer culture*, ed. David Buckingham and Vebjørg Tingstad, 80–93. Great Britain: Palgrave Macmillan.
Johnson, Daniel. 2009. 'Lingua franca'—Portal and the deconstruction of the institution. Game Set Watch. http://www.gamesetwatch.com/2009/06/column_lingua_franca_portal_an.php. Accessed 16 April 2020.
Johnson, Derek. 2013. *Media franchising: Creative license and collaboration in the culture industries*. New York: New York University Press.
Johnson, Derek. 2014a. Chicks with bricks: Building creativity across industrial design cultures and gendered construction play. In *LEGO studies*, ed. Mark J.P. Wolf, 81–104. New York: Routledge.
Johnson, Derek. 2014. Figuring identity: Media licensing and the racialization of LEGO bodies. *International Journal of Cultural Studies* 17 (4): 307–325.
Juul, Jesper. 2005. *Half-real: Video games between real rules and fictional worlds*. Cambridge: The MIT Press.
Kant, Immanuel. 1929. *Kant's inaugural dissertation and early writings on space*, trans. John Handyside. Chicago: The Open Court Publishing Company.
Kinder, Marsha. 1993. *Playing with power in movies, television, and video games: From Muppet Babies to Teenage Mutant Ninja Turtles*. Berkeley: University of California Press.
Klastrup, Lisbeth, and Susana Tosca. 2004. *International Conference on Cyberworlds* 2004: 409–416.
Kline, Stephen. 1993. *Out of the garden: Toys, TV, and children's culture in the age of marketing*. London: Verso.
Konzack, Lars. 2014. The cultural history of LEGO. In *LEGO studies*, ed. Mark J.P. Wolf, 1–14. New York: Routledge.
Kullman, Kim. 2009. Enacting traffic spaces. *Space and Culture* 12 (2): 205–217.
Kuznets, Lois Rostow. 1994. *When toys come alive: Narratives of animation, metaphorphosis, and development*. New Haven: Yale University Press.
Kylinger, Malin. 2018. Worlds inside of me. Flickr. https://www.flickr.com/photos/bluebuilder/31079405307/. Accessed 24 April 2020.
Lambie, Ryan. 2018. Greebles: How tiny details make a huge Star Wars universe. *Den of Geek*. https://www.denofgeek.com/movies/greebles-how-tiny-details-make-a-huge-star-wars-universe/. Accessed 19 April 2020.

Landay, Lori. 2014. Myth blocks: How LEGO transmedia configures and remixes mythic structures in the Ninjago and Chima themes. In *LEGO studies*, ed. Mark J.P. Wolf, 55–80. New York: Routledge.

Lauwaert, Maaike. 2009. *The place of play: Toys and digital cultures*. Amsterdam: Amsterdam University Press.

Lee, Jonathan Rey. 2014. The plastic art of LEGO: An essay into material culture. In *Design, mediation, and the posthuman*, ed. Dennis M. Weiss, Amy D. Propen, and Colby Emmerson Reid, 95–112. Lanham: Lexington Books.

Lee, Jonathan Rey. 2019. Master building and creative vision in *The LEGO Movie*. In *Cultural studies of LEGO*, ed. Rebecca C. Hains and Sharon R. Mazzarella, 149–173. Cham: Palgrave Macmillan.

Legg, Aaron. 2013. Storm-trippin' (original). YouTube. https://www.youtube.com/watch?v=7SpNHMwwID4. Accessed 19 April 2020.

Legg, Aaron. 2015. Storm Trippin 2—A new home. YouTube. https://www.youtube.com/watch?v=fTS0ez47FHE. Accessed 19 April 2020.

LEGO architecture studio [booklet]. 2013. Set 21050.

A LEGO brickumentary [Film]. 2014. Directed by Kief Davidson and Daniel Junge, Radius.

LEGO Dimensions [Videogame]. 2015. Xbox 360.

The LEGO Group. 2012. The LEGO® story. YouTube. https://www.youtube.com/watch?v=NdDU_BBJW9Y. Accessed 24 April 2020.

The LEGO Group. 2014. A short presentation. LEGO.com. https://www.lego.com/r/aboutus/-/media/about%20us/media%20assets%20library/company%20profiles/the_lego_group_a%20short%20presentation_2014_english_ed2.pdf. Accessed 7 October 2014.

The LEGO Group. 2014. Inspire imagination and keep building. YouTube. https://www.youtube.com/watch?v=BfhV3Q4LJPM. Accessed 26 September 2019.

The LEGO Group. 2015. How to Play. LEGO.com. http://www.lego.com/en-us/dimensions/how-to-play. Accessed 25 February 2016.

The LEGO Group. 2019. The greatest battle built since 1999. LEGO.com. https://www.lego.com/en-us/aboutus/news/2019/april/the-greatest-battles-built-since-1999-celebrating-20-years-of-lego-star-wars-fandom/. Accessed 19 April 2020.

LEGO Ideas Minecraft micro world—The forest [booklet]. 2012. Set 21102.

The LEGO movie [Film]. 2014. Directed by Phil Lord and Christopher Miller, Warner Bros.

The LEGO Movie 2: The second part [Film]. 2019. Directed by Mike Mitchell, Warner Bros.

LEGO *Star Wars* Hoth Wampa cave [booklet]. 2010. Set 8089.

LEGO Worlds [Videogame]. 2015. Steam. 13 September 2018 Software Update.

Lévi-Strauss, Claude. 1966. *The savage mind*. Letchworth: Weidenfeld and Nicolson Ltd.

Lipkowitz, Daniel. 2012. *The LEGO book*, 2nd ed. London: DK.

Mattes, Ari. 2019. Everything is awesome when you're part of a list: The flattening of distinction in post-ironic LEGO media. In *Cultural studies of LEGO*, ed. Rebecca C. Hains and Sharon R. Mazzarella, 73–95. Cham: Palgrave Macmillan.

McCloud, Scott. 1994. *Understanding comics: The invisible art*. New York: HarperPerennial.

McGraw, John J., Sebastian Wallot, Panagiotis Mitkidis, Andreas Roepstorff, et al. 2014. Culture's building blocks: Investigating cultural evolution in a LEGO construction task. *Frontiers in Psychology* 5: 1–12.

McKee, Robert, and Bass El-Wakil. 2014. Has *The LEGO Movie* killed the hero's journey? Wordpress. http://mckeestory.wordpress.com/2014/03/04/has-the-lego-movie-killed-the-heros-journey/. Accessed 24 June 2014.

McLuhan, Marshall. 2005. *Understanding Media*, 2nd ed. London: Routledge.

Merskin, Debra. 2019. Mia had a little lamb: Gender and species stereotypes in LEGO Sets. In *Cultural studies of LEGO*, ed. Rebecca C. Hains and Sharon R. Mazzarella, 271–295. Cham: Palgrave Macmillan.

Miller, Andrew H. 2008. *The burdens of perfection: On ethics and reading in Nineteenth-Century British Literature*. Ithaca: Cornell University Press.

Millhauser, Steven. 1983. The fascination of the miniature. *Grand Street* 2 (4): 128–135.

Moreau, C. Page, and Marit Gundersen Engeset. 2016. The downstream consequences of problem-solving mindsets: How playing with LEGO influences creativity. *Journal of Marketing Research* LIII: 18–30.

Newman, James, and Iain Simons. 2011. Using the force: *LEGO Star Wars: The Video Game*, intertextuality, narrative, and play. In *New narratives*, ed. Ruth Page and Bronwen Thomas, 239–253. Lincoln: University of Nebraska Press.

Nipper, Mads. 2012. LEGO Group commentary on attracting more girls to construction play. LEGO.com. https://www.lego.com/en-us/aboutus/news-room/2012/january/lego-group-commentary-on-attracting-more-girls-to-construction-play. Accessed 30 October 2019.

Ogata, Amy F. 2013. *Designing the creative child: Playthings and places in midcentury America*. Minneapolis: University of Minnesota Press.

Orenstein, Peggy. 2011. *Cinderella ate my daughter: Dispatches from the front lines of the new girlie-girl culture*. New York: HarperCollins.

Pickett, David. 2012. Part II: Historical perspective on the LEGO gender gap. The Society Pages. https://thesocietypages.org/socimages/2012/05/15/part-ii-historical-perspective-on-the-lego-gender-gap/. Accessed 15 April 2020.

Price, Steve. 2015. The Friends star. Flickr. https://www.flickr.com/photos/pricey73/16736883714/. Accessed 16 April 2020.

Prout, Alan, and Allison James. 2015. A new paradigm for the sociology of childhood?: Provenance, promise and problems. In *Constructing and reconstructing childhood*, ed. Allison James and Alan Prout, 6–28. New York: Routledge.

Reich, Stephanie M., Rebecca W. Black, and Tammie Foliaki. 2018. Constructing difference: Lego® set narratives promote stereotypic gender roles and play. *Sex Roles* 79: 285–298.

Robertson, David C. 2013. *Brick by brick: How LEGO rewrote the rules of innovation and conquered the global toy industry*. New York: Crown Business.

Ryan, Marie-Laure. 2017. The aesthetics of proliferation. In *World building: Transmedia, fans, industries*, ed. Marta Boni, 31–46. Amsterdam: Amsterdam University Press.

Sarkeesian, Anita. 2012. LEGO Friends—LEGO & gender part 1. YouTube. https://www.youtube.com/watch?v=CrmRxGLn0Bk. Accessed 24 April 2020.

Sawaya, Nathan. 2014. LEGO: The imperfect art tool. In *LEGO studies*, ed. Mark J.P. Wolf, 206–215. New York: Routledge.

Sawaya, Nathan. 2015. *The art of the brick: A life in LEGO*. San Francisco: No Starch Press.

Schut, Kevin. 2014. The virtualization of LEGO. In *LEGO studies*, ed. Mark J.P. Wolf, 227–240. New York: Routledge.

Seiter, Ellen. 1993. *Sold separately: Children and parents in consumer culture*. New Brunswick: Rutgers University Press.

Speaight, George. 1969. *The history of the English toy theatre*. Boston: Plays Inc.

Steinberg, Marc. 2012. *Anime's media mix: Franchising toys and characters in Japan*. Minneapolis: University of Minnesota Press

Stewart, Susan. 1984. *On longing: Narratives of the miniature, the gigantic, the souvenir, the collection*. Baltimore: The Johns Hopkins University Press.

Sutton-Smith, Brian. 1986. *Toys as culture*. New York: Gardner.

Thibault, Mattia. 2019. Transmediality and the brick: Differences and similarities between analog and digital Lego play. In *Intermedia games*, ed. Michael Fuchs and Jeff Thoss, 231–248. New York: Bloomsbury Academic.

Thita. 2011. Featured LEGO artist: Mike Doyle. *The Brick Blogger*. http://thebrickblogger.com/2011/09/mike-doyle-lego-artist/. Accessed 15 April 2020.

Toups, Zachary O., Nicole K. Crenshaw, Rina R. Wehbe, Gustavo F. Tondello, and Lennart E. Nacke. 2016. "The collecting itself feels good": Towards collection interfaces for digital game objects. *CHI PLAY '16*: 276–290.

Turkle, Sherry. 1997. *Life on the screen: Identity in the age of the internet*. New York: Touchstone.

Van Alphen, Ernst. 2001. Playing the Holocaust. In *Mirroring evil*, ed. Normal L. Kleeblatt, 65–84. New York: Rutgers University Press.

van den Bos, Jeroen, and Davy Landman. 2012. A Turing Machine built using LEGO. Legoturingmachine.org. www.legoturingmachine.org/lego-turing-machine/. Accessed 27 September 2016.

Varul, Matthias Zick. 2018. The cultural tragedy of production and the expropriation of the brickolariat: *The Lego Movie* as consumer-capitalist myth. *European Journal of Cultural Studies* 21 (6): 724–743.

Varul, Matthias Zick. 2019. The accursed second part: Small-scale discourses of gender and race in *The LEGO Movie 2*. In *Cultural studies of LEGO*, ed. Rebecca C. Hains and Sharon R. Mazzarella, 123–145. Cham: Palgrave Macmillan.

Walton, Kendall L. 1990. *Mimesis as make-believe: On the foundations of the representational arts*. Cambridge: Harvard University Press.

Warner Bros. 2015a. LEGO dimensions—Alison Brie trailer. YouTube https://www.youtube.com/watch?v=efZjpArWNZs. Accessed 4 May 2020.

Warner Bros. 2015b. LEGO dimensions—Joel McHale trailer. YouTube. https://www.youtube.com/watch?v=WXnOVyZaGK0. Accessed 4 May 2020.

Wittgenstein, Ludwig. 1922. *Tractatus logico-philosophicus*, trans. F.P. Ramsey. New York: Harcourt, Brace & Company.

Wolf, Mark J.P. 2012. *Building imaginary worlds: The theory and history of subcreation*. New York: Routledge.

Wolf, Mark J.P. 2014. Adapting the Death Star into LEGO: The case of LEGO set #10188. In *LEGO studies*, ed. Mark J.P. Wolf, 15–39. New York: Routledge.

Zinguer, Tamar. 2015. *Architecture in play: Intimations of modernism in architectural toys*. Charlottesville: University of Virginia Press.

Zosh, Jennifer M., Kathy Hirsh-Pasek, Emily J. Hopkins, Hanne Jensen, Claire Liu, S. Dave Neale, Lynneth Solis, and David Whitebread. 2018. Accessing the inaccessible: Redefining play as a spectrum. *Frontiers in Psychology* 9: 1–12.

Index

J. R. Lee, *Deconstructing LEGO*,
https://doi.org/10.1007/978-3-030-53665-7

The manufacturer's authorised representative in the EU is Springer Nature Customer Service Centre GmbH, Europaplatz 3, 69115 Heidelberg, Germany. If you have any concerns regarding our products, please contact ProductSafety@springernature.com

Printed and bound by CPI Group (UK) Ltd, Croydon, CR0 4YY

22/09/2026

02221059-0003